Health, Safety, and Environmental Management in Offshore and Petroleum Engineering

Health, Safety, and Environmental Management in Offshore and Petroleum Engineering

Srinivasan Chandrasekaran

Department of Ocean Engineering
Indian Institute of Technology Madras
India

WILEY

This edition first published 2016
© 2016 John Wiley & Sons, Ltd.
First Edition published in 2016

Registered Office
John Wiley & Sons, Ltd, The Atrium, Southern Gate, Chichester, West Sussex, PO19 8SQ,
United Kingdom

For details of our global editorial offices, for customer services and for information about how
to apply for permission to reuse the copyright material in this book please see our website at
www.wiley.com.

Library of Congress Cataloging-in-Publication Data

Names: Chandrasekaran, Srinivasan, author.
Title: Health, safety, and environmental management in offshore and petroleum
    engineering / Srinivasan Chandrasekaran.
Description: First edition. | Chichester, West Sussex, United Kingdom :
    John Wiley & Sons, Ltd., [2016] | Includes bibliographical references and index.
Identifiers: LCCN 2015047419 (print) | LCCN 2015049980 (ebook) |
    ISBN 9781119221845 (cloth) | ISBN 9781119221425 (pdf) | ISBN 9781119221432 (epub)
Subjects: LCSH: Offshore structures–Safety measures. | Offshore structures–Risk assessment. |
    Petroleum engineering–Safety measures. | Petroleum engineering–Risk assessment. |
    Petroleum in submerged lands–Environmental aspects. | Natural gas in submerged
    lands–Environmental aspects.
Classification: LCC TC1665 .C457 2016 (print) | LCC TC1665 (ebook) | DDC 622/.8–dc23
LC record available at http://lccn.loc.gov/2015047419

A catalogue record for this book is available from the British Library.

Set in 10/12.5pt Palatino by SPi Global, Pondicherry, India
Printed and bound in Singapore by Markono Print Media Pte Ltd

1   2016

Contents

Preface

The regulations of risks to health, safety, and environmental management that arise from the exploration and production works in the oil and gas industries are gaining more attention in the recent past. There is a growing necessity to maintain good and healthy work-space for people on board and also to protect the fragile ecosystem. The unregulated use of chemicals or other hazardous substances in oil and gas industries can challenge the technical workforce by putting their health at risk, causing various levels of discomfort in addition to causing catastrophic damage to the offshore assets. Accidents reported in the recent past in oil and gas sector also demonstrate the seriousness of Health, Safety, and Environmental Management in this domain of workspace. The objective of the book is to share the technical know-how in the field of health, safety, and environmental management, applicable to oil and gas industries. Contents of the book are spread across four chapters, addressing the vital areas of interest in HSE, as applicable to offshore and petroleum engineering. The first chapter highlights safety assurance and assessment, emphasizing the need for safety. The second chapter focuses on the environmental issues and management that arise from oil and gas exploration. The third chapter deals with the accident modeling, risk assessment, and management, while the fourth chapter is focused on safety measures in design and operations. The book explains the concepts in HSE through a simple and straightforward approach, which makes it comfortable for practicing engineers as well. The focus however is capacity building in safety and risk assessment, which is achieved through a variety of example problems and case studies. The author's experiences in both the academia and leading oil and gas industries are shared through the illustrated case studies. The book is an important milestone in the capacity building of young engineers and preparing them for a safe exploration process. Sincere thanks are due to Centre for Continuing Education, IIT Madras for assisting in writing this book.

Srinivasan Chandrasekaran

About the Author

Professor Srinivasan Chandrasekaran is a Professor in the Department of Ocean Engineering, Indian Institute of Technology, Madras, India. He has teaching, research, and industrial experience of about 24 years during which he has supervised many sponsored research projects and offshore consultancy assignments both in India and abroad. His active areas of research include dynamic analysis and design of offshore platforms, development of geometric forms of compliant offshore structures for ultra-deep water oil exploration and production, structural health monitoring of ocean structures, seismic analysis, and design of structures and risk analyses and reliability studies of offshore and petroleum engineering plants. He was a visiting fellow under the invitation of Ministry of Italian University Research to University of Naples Federico II, Italy, for a period of 2 years during which he conducted research on advanced nonlinear modeling and analysis of structures under different environmental loads with experimental verifications. He has published about 140 research papers in international journals and refereed conferences organized by professional societies around the world. He has authored five textbooks, which are quite popular among the graduate students of civil and ocean engineering: *Seismic Design Aids for Nonlinear Analysis of Reinforced Concrete Structures* (ISBN: 978-1-4398-0914-3); *Analysis and Design of Offshore Structures with Illustrated Examples* (ISBN: 978-89-963915-5-5); *Advanced Theory on Offshore Plant FEED Engineering* (ISBN: 978-89-969792-8-9); *Dynamic Analysis and Design of Offshore Structures* (ISBN: 978-81-322-2276-7); *Advanced Marine Structures* (ISBN: 978-14-987-3968-9). His books are also recommended as reference material in many universities in India and abroad.

He also conducted two online courses under Mass Open Online Courses (MOOC) under NPTEL, GoI titled Dynamic analysis of offshore structures and HSE in oil offshore and petroleum industries. He is a member of many national and international professional bodies and has delivered many invited lectures and keynote addresses in the international conferences, workshops, and seminars organized in India and abroad. He has also delivered four web-based courses:

• Dynamic Analysis of Ocean Structures (http://nptel.ac.in/courses/114106036/)
• Ocean Structures and Materials (http://nptel.ac.in/courses/114106035/)
• Advanced Marine Structures (http://nptel.ac.in/courses/114106037/)
• Health, Safety and Management in Offshore and Petroleum Engineering (http://nptel.ac.in/courses/114106017/)

under the auspices of National Program on Technology Enhancement Learning (NPTEL), Government of India.

1

Safety Assurance and Assessment

Introduction to Safety, Health, and Environment Management

Health, Safety, and Environmental (HSE) management is an integral part of any business and is considered to be extremely essential when it comes to managing business in oil and gas sectors. HSE requirements are generally laid out considering the expectations of the divisional compliance with that of the standard policies. This is the most important part of HSE through legislation in the recent decades and thus forms the basis of HSE regulations in the present era. Apart from setting out the general duties and responsibilities of the employers and others, it also lays the foundation for subsequent legislation, regulations, and enforcement regimes. HSE standards are circumscribed around activities that are "reasonably practicable" to assure safety of the employees and assets as well. HSE regulations impose general duties on employers for facilitating the employees with minimum health and safety norms and members of the public; general duties on employees for their own health and safety and that of other employees, which are insisted as regulations.

Health, Safety, and Environmental Management in Offshore and Petroleum Engineering, First Edition.
Srinivasan Chandrasekaran.
© 2016 John Wiley & Sons, Ltd. Published 2016 by John Wiley & Sons, Ltd.
Companion website: www.wiley.com/go/chandrasekaran/hse

1.1 Importance of Safety

There are risks associated with every kind of work and workplace in day-to-day life. Levels of risk involved in some industries may be higher or lower due to the consequences involved. These consequences affect the industry as well as the society, which may create a negative impact on the market depending upon the level of risk involved (Ale, 2002). It is therefore very important to prevent death or injury to workers, general public, prevent physical and financial loss to the plant, prevent damage to the third party, and to the environment. Hence, rules and regulations for assuring safety are framed and strictly enforced in offshore and petroleum industries, which is considered to be one of the most hazardous industries (Arshad Ayub, 2011). The prime goal is to protect the public, property, and environment in which they work and live. It is a commitment for all industries and other stakeholders toward the interests of customers, employees, and others. One of the major objectives of the oil and gas industries is to carry out the intended operations without injuries or damage to equipment or the environment. Industries need to form rules, which will include all applicable laws and relevant industry standards of practice. Industries need to continuously evaluate the HSE aspects of equipment and services. It is important for oil and gas industries to believe that effective HSE management will ensure a good business. Continuous improvement in HSE management practices will yield good return in the business apart from ensuring goodness of the employees (Bottelberghs, 2000). From the top management through the entry level, every employee should feel responsible and accountable for HSE. Industries need to be committed to the integration of HSE objectives into management systems at all levels. This will not only enhance the business, but also increase the success rate by reducing risk and adding value to the customer services.

1.2 Basic Terminologies in HSE

ALARP: To reduce a risk to a level 'as low as reasonably practical' (ALARP). It involves balancing reduction in risk against time, trouble, difficulty, and cost of achieving it. Cost of further reduction measures become unreasonably disproportionate to the additional risk reduction obtained.

Audit: A systematic, independent evaluation to determine whether or not the HSE-MS and its operations comply with planned arrangements. It also examines whether system is implemented effectively and is suitable to fulfill the company's HSE policies and objectives.

Client: A company that issues a contract to a contractor or subcontractor. In this document the client will generally be an oil and gas exploration company that will issue a contract to a contractor to carry out the work. The contractor may then take the role of a client by issuing contract(s) to subcontractor(s).

Contract(s): An agreement between two parties in which both are bound by law and which can therefore be enforced in a court or other equivalent forum.

Contractor(s): An individual or a company carrying out work under a written or verbally agreed contract for a client.

Hazard: An object, physical effect, or condition with the potential to harm people, the environment, or property.

HSE: Health, safety, and environment. This is a set of guidelines, in which security and social responsibilities are recognized as integral elements of HSE management system.

HSE capability assessment: A method of screening potential contractors to establish that they have the necessary experience and capability to undertake the assigned work in a responsible manner while knowing how to effectively deal with the associated risks.

HSE Plan: Is a definitive plan, including any interface topics, which sets out the complete system of HSE management for a particular contract.

Incident: An event or chain of events that has caused or could have caused injury or illness to people and/or damage (loss) to the environment, assets, or third parties. It includes near-miss events also.

Inspection: A system of checking that an operating system is in place and is working satisfactorily. Usually this is conducted by a manager and with the aid of a prepared checklists. It is important to note that this is not the same as an audit.

Interface: A documented identification of relevant gaps (including roles, responsibilities, and actions) in the different HSE-MS of the participating parties in a contract, which, when added to the HSE plan will combine to provide an operating system to manage all HSE aspects encountered in the contract with maximum efficiency and effectiveness.

Leading indicator: A measure that, if adopted, helps to improve performance.

Subcontractor(s): An individual or company performing some of the work within a contract, and under contract to either the original client or contractor.

Third party: Individuals, groups of people, or companies, other than the principal contracted parties, that may be affected by or involved with the contract.

Toolbox meeting: A meeting held by the workforce at the workplace to discuss HSE hazards that may be encountered during work and the procedures that are in place to successfully manage these hazards. Usually this is held at the start of the day's work; a process of continual awareness and improvement.

Accident: It refers to the occurrence of single or sequence of events that produce unintended loss. It refers to the occurrence of events only and not the magnitude of events.

Safety or loss Prevention: It is the prevention of hazard occurrence (accidents) through proper hazard identification, assessment, and elimination.

Consequence: It is the measure of expected effects on the results of an incident.

Risk: It is the measure of the magnitude of damage along with its probability of occurrence. In other words, it is the product of the chance that a specific undesired event will occur and the severity of the consequences of the event.

Risk analysis: It is the quantitative estimate of risk using engineering evaluation and mathematical techniques. It involves estimation of hazard, their probability of occurrence, and a combination of both.

Hazard analysis: It is the identification of undesired events that lead to materialization of a hazard. It includes analysis of the mechanisms by which these undesired events could occur and estimation of the extent, magnitude, and likelihood of any harmful effects.

Safety program: Good program identifies and eliminates existing safety hazards. Outstanding program prevents the existence of a hazard in the first place. Ingredients of a safety program are safety knowledge, safety experience, technical competence, safety management support, and commitment to safety.

Initial response from HSE: There are two sets of regimes namely: (i) goal- setting regimes; and (ii) rule- based regimes. *Goal-setting regimes* have a duty holder who assesses the risk. They should demonstrate its understanding and controls the management, technical, and systems issues. They should keep pace with new knowledge and should give an opportunity for workforce involvement. *Rule-based regimes* consist of a legislator who sets the rules. They emphasizes compliance rather than outcomes. The disadvantage is that they it are slow to respond. They gives less emphasis on continuous improvement and less work force involvement.

1.2.1 What Is Safety?

Safety is a healthy activity of prevention from being exposed to hazardous situation. By remaining safe, the disastrous consequences are avoided, thereby saving the life of human and plant in the industry.

1.2.2 Why Is Safety Important?

Any living creature around the world prefers to be safe rather than risk themselves to unfavorable conditions. The term safety is always associated with risk. When the chances of risks are higher then the situation is said to be highly unsafe. Therefore, risk has to be assessed and eliminated and safety has to be assured.

1.3 Importance of Safety in Offshore and Petroleum Industries

Safety assurance is important in offshore and petroleum industries as they are highly prone to hazardous situations. Two good reasons for practicing safety are: (i) investment in an offshore industry is several times higher than that of any other process/production industry across the world and (ii) offshore platform designs are very complex and innovative and hence it is not easy to reconstruct the design if any damage occurs (Bhattacharyya et al., 2010a, b). Prior to analyzing the importance of safety in offshore industries, one should understand the key issues in petroleum processing and production. Safety can be ensured by identifying and assessing the hazards in each and every stages of operation. Identification and assessment of hazard at every stages of operation are vital for monitoring safety, both in quantitative and qualitative terms. Prime importance of safety is to ensure prevention of death or injury to workers in the plant and also to the public located around. Safety should also be checked in terms of financial damage to the plant as investment is huge in oil and petroleum industries than any other industry. Safety must be ensured in such a way that the surrounding atmosphere is not contaminated (Brazier and Greenwood, 1998).

Piper Alpha suffered an explosion on July 1988, which is still regarded as one of the worst offshore oil disasters in the history of the United Kingdom (Figure 1.1). About 165 persons lost their lives along with 220 crew members. The accident is attributed mainly due to a human error and is a major eye-opener for the offshore industry to revisit safety issues. Estimation of property damage is about $1.4 billion. It is understood that the accident was

Figure 1.1 Piper Alpha disaster

mainly caused by negligence. Maintenance work was simultaneously carried out in one of the high-pressure condensate pumps' safety valve, which led to the leak of condensates and that resulted in the accident. After the removal of one of the gas condensate pumps' pressure safety valve for maintenance, the condensate pipe remained temporarily sealed with a blind flange as the work was not completed during the day shift. The night crew, who were unaware of the maintenance work being carried out in the last shift on one of the pumps, turned on the alternate pump. Following this, the blind flange, including firewalls, failed to handle the pressure, leading to several explosions. Intensified fire exploded due to the failure in closing the flow of gas from the Tartan Platform. Automatic fire fighting system remained inactive since divers worked underwater before the incident. One could therefore infer that the source of this devastating incident was due to a human error and lack of training in shift-handovers. Post this incident, significant (and stringent) changes were brought in the offshore industry with regard to safety management, regulation, and training (Kiran, 2014).

On March 23, 1989, Exxon Valdez, which was on its way from Valdez, Alaska, with a cargo of 180 000 tons of crude oil collided with an iceberg and 11 cargo tanks, got punctured. Within a few hours 19 000 tons of crude oil was lost. By the time the tanker was refloated on April 5, 1989, about 37 000 tons was lost. In addition, about 6600 km^2 of the country's greatest fishing grounds and the surrounding shoreline were sheathed in oil. The size of the

Figure 1.2　Exxon Valdez oil spill

spill and its remote location made it difficult for the government and industry to salvage the situation. This spill was about 20% of the 18 000 tons of crude oil, which the vessel was carrying when it struck the reef (Figure 1.2).

Safety plays a very important role in the offshore industry. Safety can be achieved by adopting and implementing control methods such as regular monitoring of temperature and pressure inside the plant, by means of well-equipped coolant system, proper functioning of check valves and vent outs, effective casing or shielding of the system and check for oil spillages into the water bodies, by thoroughly ensuring proper control facilities one can avoid or minimize the hazardous environment in the offshore industry (Chandrasekaran, 2011a, b).

1.4 Objectives of HSE

The overall objective is to describe a process by which clients can select suitable contractors and award contracts with a view to improving the client and contractor management on HSE performance in upstream activities. For brevity, security, and social responsibilities have not been included in the document title; however, they are recognized as integral elements of the

HSE-management systems. Active and ongoing participation by the client, contractor, and their subcontractors are essential to achieve the goal of effective HSE management. While each has a distinct role to play in ensuring the ongoing safety of all involved, there is an opportunity to further enhance the client–contractor relationship by clearly defining roles and responsibilities, establishing attainable objectives, and maintaining communication throughout the contract lifecycle. The aims of HSE practice are to improve performance by:

- Providing an effective management of HSE in a contract environment, so that both the client and the contractor can devote their resources to improve HSE performance.
- Facilitating the interface of the contractor's activities with those of the client, other contractors, and subcontractors so that HSE becomes an integrated activity of all facets of process.

These guidelines are generally formulated and provided to assist clients, contractors, and subcontractors to clarify the process of managing HSE in contract operations (Chandrasekaran, 2014a, b). This generated document does not replace the necessary professional judgment needed to recommend the specific contracting strategy to be followed. Each reader should analyze his or her particular situation and then modify the information provided in this document to meet their specific needs to obtain appropriate technical support wherever required. Oil and Gas Production Secretariat is the custodian of these guidelines and will initiate updates and modifications based upon review and feedback from users through periodic meetings. In general, these guidelines are not intended to take precedence over a host country's legal or other requirements (Chandrasekaran, 2011e).

1.5 Scope of HSE Guidelines

HSE guidelines provide a framework for developing and managing contracts in offshore industry. While HSE aspects are important in the development of a contract strategy, these guidelines do not cover many vital aspects of the contract process. They prescribe various phases of the contracting process and associated responsibilities of the client, contractors, and subcontractors. It begins with planning and ends with evaluation of the contract process.

1.6 Need for Safety

Employers establish teams, such as quality assurance (or control) teams to get employees involved in the quality process. Employees are empowered to stop an entire production line if they become aware of any problem affecting production or quality. This is a common industrial practice as this ensures increased participation for improving quality standards and also to reduce the cost line. A similar trend is necessary in practicing safety norms as well. Unfortunately, it is observed that in many process industries, employees are not involved in the safety process except that they are members of the safety committee. But it is important to realize that if one desires to improve something for which employees are responsible, then one should establish it as an important component of their workday by making it an important element of their business. By involving the employees in the safety assurance program, they get a keen sensation of consciousness and ownership; results include better production and lower price. It is not recommended to punish a worker who broke a safety principle but turn a blind eye to the supervisor or manager who sanctioned the violation through his/her silence. The task of the supervisor or manager is to guarantee that the job is performed right and safe.

As Managers are part of the system that challenges safety, they should also be responsible to provide the answer to the perceived challenges. Long-lasting safety success cannot be assured unless the management team is a function of the safety effort. The goal of every organization should be to build a safety culture through employee engagement. By getting employees involved in performing inspections, investigations, and other procedures, needs of safety and health programs can be easily met. Employee safety can be maximized by making safety culture through increased consciousness. In particular, a skillful director of an oil company will make every effort to improve and regularize the outcome of the business in its entirety, although it is not unusual for a manager to excel in certain fields. In the workplace, there are several micro issues that must be successfully managed for the company to succeed in the business. One may establish quotas or reward individual achievements to recognize outstanding production effort of an individual employee or a group of employees. Alternatively, one should ensure that in this rigorous task, safety in not compromised even unknowingly. As for safety and health, if the company contrives to manage them for the maximum success, then there is also a need to execute the program in the same manner. Safety managers are the experts who coordinate efforts and keep top management informed on issues linked to safety and health.

Policies and procedures, along with the signs and warnings, provide some measures of restraint. The point of control is only as effective as the level of enforcement of the indemnities. Where enforcement is weak, control and thus compliance are weak as well. The best-suited example is the signboard, which is utilized as a way of mastering the speed point of accumulation in highways. But only where the signs are strictly enforced can one can see the drivers complying with the indicated speed limits. In most of the cases, they will drive as fast as they think law enforcement will take into account. Therefore, it is not the signal that controls speed on the highway; it is the degree of enforcement established by local law. Therefore, to prevent employee injury and sickness, one should maximize the management of safety and health at workplaces.

1.7 Organizing Safety

Major accidents reported in oil and gas industries in the past are important sources of information for understanding safety. Lessons learnt from these accidents, through detailed diagnosis, will be helpful in preventing the occurrence of similar accidents in the future. It is evident from the literature that in the last 15 years, major accidents in the offshore industry has declined (Khan and Abbasi, 1999). It is true that the important experiences gained from these events may be blanked out and the information may not be brought forward to the future generation if analyses of such accidents are not reported. The major risk groups in offshore and oil industry are blowouts, hydrocarbon leaks on installations, hydrocarbon leaks from pipelines/risers, and structural failures (Vinnem, 2007a). Some of the major accidents that took place in the past and the lessons learnt from these accidents are discussed in the next section.

1.7.1 Ekofisk B Blowout

On April 23, 1977, a blowout occurred in the steel jacket wellhead platform during a workover on a production well. The Blow Out Preventer (BOP) was not in place and could not be reassembled on demand. All the personnel on board were rescued, through the supply vessel, without injuries but the accident resulted in the oil spill of about $20\,000\,m^3$. The well was then mechanically capped after 7 days after the event and production was shut down for half a dozen weeks to allow cleanup operations. Although the Ekofisk B blowout did not result in any human death or material damage and was

Figure 1.3 Ekofisk blowout

exclusively limited to spills, an important lesson learnt is that capping of a blowout is possible, although it requires time. This may be vital information from a design point of view, which can be considered in modeling and analysis of BOPs (Kiran, 2012) (Figure 1.3).

1.7.2 Enchova Blowout

On August 16, 1984, a blowout occurred on the Brazilian fixed jacket platform Enchova-1. It was producing 40 000 barrels of oil and 1 500 000 m³ of gas per day through 10 wells. The first fire was due to ignition of gas released during drilling, which was under constraint. But, the fire due to oil leakage led to a knock. The ensuing flame was blown out late the following day. The platform's drilling equipment was gutted but the remainder of the platform remained intact. Thirty-six people were killed while evacuating as the lifeboat malfunctioned, 207 survivors were rescued from the platform through helicopters and lifeboats. The most vital lesson learnt from the accident was the use of conventional lifeboats for evacuation purposes. Failure of hooks in the lifeboat gained attention and led to improvement in the design later on. Lack of competence to control the release mechanism led to stringent training of personnel on safety operations during rescue and emergency situations (Chandrasekaran, 2011d) (Figure 1.4).

Figure 1.4   Enchova blowout

1.7.3 West Vanguard Gas Blowout

The semisubmersible drilling unit, West Vanguard, experienced a gas blow-out on October 6, 1985, while conducting exploration drilling in the Haltenbanken area, Norway. During drilling, the drill bit entered a thin gas layer, which was about 236 m below the sea bottom. This caused an influx of gas into the wellbore, which was followed by a second influx of gas after a day; third influx of gas had a gas blowout. It was noticed that the drilling operation was carried out without the use of BOP. When the drilling crew realized the gas blow out happened, inexperienced personnel started pumping heavy mud and also opened the valve to divert gas flow away from the drill stack. But, within minutes, erosion in the bends of the diverter caused the escape and the gas entered the cellar deck from the bottom. An attempt to release the coupling of the well head of the marine riser, located on the sea bed, was unsuccessful due to the ignition hazard in all areas of the platform. Ignition finally occurred from the engine room in 20 minutes after the initial start of the event, which led to a strong explosion and a fire. Two lifeboats were launched for the crew members immediately after the burst. One of the

Figure 1.5   West Vanguard gas blowout

lessons learnt was the time management of launching lifeboats, which saved the lives of people onboard. However, inexperienced attempts made to divert the gas flow away from the drilling stack remained an important lesson to learn (Figure 1.5).

1.7.4 Ekofisk A Riser Rupture

The riser of steel jacket wellhead platform Ekofisk Alpha ruptured due to fatigue failure on November 1, 1975. The failure occurred due to insufficient protection in the splash zone and led to rapid corrosion. Leaks occurred at once at a lower part of the living quarters, causing an explosion and flame propagation. Intense flame remained for a short duration as the gas flow was immediately shut down; the blast was completely eliminated within 2 hours due to the efficient design of fire-fighting system. Only a modest damage to the platform was caused due to fire. The most important lesson learnt from the accident is about the location of riser below the living quarters (Chandrasekaran, 2010b). Best training and emergency evacuation procedures adopted and practiced by the crew resulted in minor injuries with no fatalities. The platform only suffered limited fire damage due to the short duration of intensive fire loads.

1.7.5 Piper A Explosion and Fire

On July 6, 1988, an ignition caused a gas leak from the blind flange in the gas compression area of Piper A. The explosion load was estimated to be about 0.3–0.4 bar over pressure. The first riser rupture occurred after 20 minutes, from which the fire increased dramatically; this resulted in further riser ruptures. The personnel escaped from the initial explosion gathered in the accommodation and were not given any further instruction about the escape and evacuation plans. Onboard communication became nonfunctional due to initial stages of the accident. Evacuation with the aid of helicopters was not possible due to blast and smoke around the platform. A total of 166 crew members died in the incident. Most of the fatalities were due to the smoke inhalation inside the accommodation, which subsequently collapsed into ocean. From a design perspective, location of the central room, radio room, and accommodation, which were very close to the gas compression area, the accident could have been avoided (Chandrasekaran, 2015). Further, not protecting them from blast and fire barriers was also a design fault. Location of accommodation on the upside of the installation led to quick accumulation of smoke within the quarters, which is also a major design fault. Lessons learnt from the operational aspects are as follows: fire water pump was not kept on automatic standby for a long time. This was a serious failure of the installation, which led to the unavailability of water for cooling oil fire.

1.7.5.1 What Do These Events Teach Us?

From these accident cases it is well known that there is limitation of knowledge in forecasting the consequences of such incidents. Past experiences alone are not sufficient to calculate the sequence of outcomes (Kletz, 2003). This is due to the fact that such accidents are very uncommon and cannot be predicted. However, catastrophic consequences in most of the cases could have been avoided by taking proper care during the design stage and also by imparting emergency evacuation training to all personnel onboard.

1.8 Risk

Fatality and damages caused to the human and material property will result in a financial loss to the investor. Risk involves avoidance of loss and undesirable consequences. Risk involves probability and estimate of potential losses as well. According to ISO 2002, risk is defined as the *combination of probability of an event and its outcome*. ISO 13702 defines risk as *probability*

Figure 1.6 Piper Alpha explosion

at which a specified hazardous event will occur and the harshness of the effects of the case. Mathematically risk (R) can be expressed for each accident sequence i as below:

$$R = \sum_i (p_i C_i) \tag{1.1}$$

where, p is the probability of accidents and C is the consequence. The above expression gives a statistical look to the risk definition, which often means that the value in practice shall never be discovered. If the accident rates are rare, an average value will have to be assumed over a long period, with low annual values. If during 50 years, one has reported only about six major accidents with a sum of 10 fatalities, then this amounts to about 0.2 fatalities per year. Risk, therefore converts an experience into a mathematical term by attaching the consequences of the occurred events. Risk, is a post-evaluation of any event or incident, but risk can also be predicted with appropriate statistical tools (Chandrasekaran and Kiran, 2014a, b) (Figure 1.6).

1.9 Safety Assurance and Assessment

Safety and risk are contemporary. Safety is a subjective term, whereas risk is an abstract term. As safety cannot be quantified directly, it is always addressed indirectly using risk estimates. Risk can be classified into individual risk and societal risk. Individual risk is defined as the frequency at which an individual may be expected to sustain a given level of harm from the realization of hazard. It usually accounts only for the risk of death and is expressed as risk per year or Fatality Accident Rate (FAR). It is given by:

$$\text{Average individual risk} = \frac{\text{number of fatalities}}{\text{number of people at risk}} \qquad (1.2)$$

Societal risk is defined as the relationship between the frequency and number of people suffering a given level of harm from realization of any hazard. It is generally expressed as FN curves, which shows the relationship between the cumulative frequency (F) and the number of fatalities (N). It can also be expressed in annual fatality rate in which the frequency and fatality data are combined into a single convenient measure of group. As it becomes important to quantify risk, risk estimates are attractive only because of the consequences associated with the term. But for the consequences, risk remains as a mere statistical number. Now, one is interested to know methods to estimate loss. This is due to the fact that financial implications that arise from the consequences can be easily reflected in the company's balance sheet. Unfortunately, there is no single method, which is capable of measuring accident and loss statistics with respect to all required aspects. Three systems are commonly used in offshore industry, they are:

1. Occupational Safety and Health Administration, US Department of Labor (OSHA)
2. Fatal Accident Rate (FAR)
3. Fatality rate or deaths per person per year

All the methods report the number of accidents and/or fatalities for a fixed number of working hours during a specified period, which is unique and common among them (Chandrasekaran, 2015).

1.10 Frank and Morgan Logical Risk Analysis

Frank and Morgan (1979) proposed a systematic method of financing risk and presented a scheme for risk reduction. Their model is applicable to any process industries and therefore valid for oil and gas industries as well. Before applying this method for targeting risk reduction, the whole company is subdivided into several departments. This division can be based on the functional aspects or administerial aspects. This method involves six steps of risk analysis, which are as follows:

Step 1: Compute risk index for each department
Each department inherently has a risk level, which is to be identified first. This can be done by evaluating the hazards present and the control measures available. This is also called as the first level of risk assessment. It is generally done by preparing a checklist, shown in Table 1.1. Control scores and hazard scores for all the departments are established from the checklist given in Table 1.2.

Hazard checklist has six groups of hazards. There are scores associated with each hazard, within each group. These scores are summed up for hazards applied within that group. The hazard score for a group is given by:

$$\text{Hazard score} = \text{sum} \times \text{hazard weightage} \qquad (1.3)$$

Hazard score for each department is the sum of the scores computed for each of the six groups. Similarly one can estimate the control scores as well. Control score for each department is the sum of the scores of each of the six groups as tabulated above. Control score for a group is given by:

$$\text{Control score} = \text{sum} \times \text{control measure weightage} \qquad (1.4)$$

After determining the hazard and control scores for each department, risk index can be calculated as given below. Risk index may be either positive or negative depending upon the control measures and hazard groups present in each department.

$$\text{Risk index} = \text{control score} - \text{hazard score} \qquad (1.5)$$

Step 2: Determine relative risk for each department
The aim is to rank the departments and not the individual hazards present in the plant. This is due to the fact that the department with the highest risk index (highest positive value) is not likely to need much reduction in hazards. High risk index means that the controls are very effective. Those departments will need funds lesser than other department to mitigate/eliminate/reduce hazards. In fact, use the best department risk score as the base reference. All curves are normalized with respect to the best department. This is done by subtracting the risk score of the best department from risk scores of the concerned department. This adjustment will make the relative risk of best department as zero.

Table 1.1   Hazard groups and hazard score

Rating points	Hazard group and hazard (Group hazard factor in parentheses)

Fire/explosion potential (10)

2	Large inventory of flammables
2	Flammables generally distributed in the department rather than localized
2	Flammables normally in vapor phase rather than liquid phase
2	Systems opened routinely, allowing flammable/air mix, versus a totally closed system
1	Flammables having low flash points and high sensitivities
1	Flammables heated and processed above flash point

Complexity of process (8)

2	Need for precise reactant addition and control
2	Considerable instrumentation requiring special operator understanding
2	Troubleshooting by supervisor rather than operator
1	Large number of operations and/or equipment monitored by one operator
1	Complex layout of equipment and many control stations
1	Difficult to startup or shutdown operations
1	Many critical operations to be maintained

Stability of process (7)

3	Severity of uncontrolled situation
2	Materials that are sensitive to air, shock, heat, water, or other natural contaminants in the process.
2	Potential exists for uncontrolled reactions
1	Raw materials and finished goods that require special storage attention
1	Intermediates that are thermally unstable
1	Obnoxious gases present or stored under pressure

Operating pressure involved (6)

3	Process pressure in excess of $110\,lb/in^2$ (gauge)
2	Process pressure above atmosphere but less than $110\,lb/in^2$ (gauge)
1	Process pressure ranges from vacuum to atmospheric
3	Pressures are process rather than utility related
2	High pressure situations are in operator
1	Excessive sight glass application
1	Nonmetallic materials of construction in pressure service

Personnel/environment hazard potential (4)

3	Exposure to process materials pose high potential for severe burn or severe health risks

Table 1.1   (*Continued*)

Rating points	Hazard group and hazard (Group hazard factor in parentheses)
2	Process materials corrosive to equipment
2	Potential for excursion above Threshold Limit Value (TLV)
1	Spills and/or flumes have high impact on equipment, people, or services
1	High noise levels make communication difficult

High temperatures (2)

1	Equipment temperatures exist in <100°C range (low)
2	Equipment temperatures exists in 100<170°C range
3	Equipment temperatures exists in 170<230°C range
2	High temperature situations are m operator-frequented area
2	Overflows and/or leaks are fairly common
2	Heat stress possibilities from nature of work or ambient air

Step 3: Compute percentage risk index for each department
This indicates relative contribution of each department to the total risk of the plant. Relative risk of each department is converted to a percent of total risk by a simple procedure. Total risk of all departments is the sum of absolute value of relative risk of each department. The percent risk index is given by:

$$\% \text{ Risk index} = \frac{\text{relative risk}_i}{\sum\limits_{0}^{i} \text{relative risk}} \times 100 \qquad (1.6)$$

Step 4: Determine composite exposure dollars for each department
The estimated risk is subsequently converted to financial value now. This estimates the financial value of risk for each department. Composite exposure dollars are the sum of monetary value of three components: (i) property value; (ii) business interruption; and (iii) personnel exposure. Property value is estimated by the replacement cost of all materials and equipments at risk in the department. Business interruption is computed as the product of unit cost of goods and production per year and expected percentage capacity. Personnel exposure is the product of total number of people in the department during the most populated shift and the monetary value of each person.

Table 1.2   Control scores and control group

Rating points	Control group and control (Group control factor in parentheses)

Fire protection (10)

4	Automatic sprinkler system capable of meeting demands
2	Supervisors and operators knowledgeable in installed fire protection systems are trained properly response to fire
1	Adequate distribution of fire extinguishers
1	Fire protection system inspected and tested with regular frequency
1	Building and equipment provided with capability to isolate and control fire
1	Special fire detection and protection provided where indicated

Electrical integrity (8)

3	Electrical equipment installed to meet National Electrical Code area classification
1	Electrical switches labeled to identify equipment served
1	Integrity of installed electrical equipment maintained
1	Class I, division 2 installations provided with sealed devices Explosion proof equipment provided or purged reliably and good electrical isolation between hazardous and non hazardous areas.
1	All electrical equipment capable of being locked out
1	Disconnects provided, identified, inspected, and tested regularly
1	Lighting securely installed and facilities properly grounded

Safety devices (7)

3	Relief devices provided and relieving is to a safe area
2	Confidence that interlocks and alarms are operable
2	Operating instructions are complete and current, and department has continued training and/or retaining program
1	Safety devices are properly selected to match application
1	Critical safety devices identified and included in regular testing program
1	Fail safe instrumentation provided

Inerting and dip piping (5)

2	Vessels handling flammables provided with dip pipes
2	Vessels handling flammables provided with reliable inerting system
2	Effectiveness of inerting assured by regular inspection and testing
1	Inerting instruction provided and understood
1	Inerting system designed to cover routine and emergency startup
1	Equipment ground visible and tested regularly
1	Friction hot spots identified and monitored

Table 1.2   (*Continued*)

Rating points	Control group and control (Group control factor in parentheses)

Ventilation/Open construction (4)

3	No flammables exist or open air construction is provided
2	Local ventilation provided to prevent unsafe levels of flammable, toxic, or obnoxious vapors
2	Provision made for containing and controlling large spills and leaks of hazardous materials
1	Building design provides for natural ventilation to prevent accumulation of dangerous vapors
1	Sumps, pits, etc., nonexistent or else properly ventilated or monitored
1	Equipment entry prohibited until safe atmosphere assured

Accessibility and/or separation (2)

2	Critical shutdown devices and/or switches visible and accessible
2	Adjacent operations or services protected from exposure resulting from incident in concerned facility
2	Operating personnel protected from hazards by location
1	Orderly spacing of equipment and materials within the concerned facility
1	Adjacent operations offer no hazard or exposure
1	Hazardous operations within the facility well isolated

Step 5: Compute composite risk for each department

For each department, composite risk is the product of composite exposure dollars and percentage risk index of that department. This value represents the relative risk of each department. Units for composite risk are in dollars. Composite risk for each department is given by:

$$\text{Composite risk} = (\text{composite exposure}) \times (\% \text{ risk index}) \qquad (1.7)$$

Step 6: Risk ranking

This is the final step in the process. Risk ranking of the departments is done based on the composite risk as this will help the risk managers to decide the requirement of fund for each department either to mitigate risk or at least to control risk. Departments should be ranked from highest composite score to the lowest.

Table 1.3 Data for each department of the process plant

Exposure dept.	Hazard score	Control score	Property value ($) ($\times 10^3$)	Business interruption cost ($) ($\times 10^3$)	Composite score ($)	
					Personnel	Exposure dollars
A	257	304	2900	1400	900	5200
B	71	239	890	1200	653	2743
C	181	180	1700	720	1610	4030
D	152	156	290	418	642	1350
E	156	142	520	890	460	1870
F	113	336	2910	3100	1860	7870

Example problem

Now, let us consider an example to understand the application of Frank and Morgan risk analysis. Relevant data for each department is given in Table 1.3.

From the given input data, risk index is calculated using the Equation 1.5. For example, risk index of department A is given by:

$$\text{Risk index} = \text{control score} - \text{hazard score} \qquad (1.8)$$

$$\text{Risk index}_A = 304 - 257 = 47$$

Similarly, risk index for all other departments are computed. For determining the relative risk, department risk index is subtracted from the maximum risk index. In this example, maximum risk index is for department F (223), which is considered as the *reference department*. Therefore, relative risk for department A is given by:

$$\text{Relative risk}_A = 47 - 223 = -176$$

The % risk index is then calculated for all the departments as:

$$\% \text{Risk index}_A = \frac{-176}{911} \times 100 = -19.31\%$$

After computing the % risk index for each department, composite risk is calculated:

$$\text{Composite risk}_A = 5200 \times 19.31\% = 1005$$

Table 1.4 Computation of risk ranking

Exposure dept.	Risk index	Relative risk	% Risk index	Composite exposure (×10³)	Composite risk (×10³)	Risk ranking
A	47	−176	−19.31	5200	1005	1
B	168	−55	−6.04	2743	166	5
C	−1	−224	−24.59	4030	991	2
D	4	−219	−24.04	1350	325	4
E	−14	−237	−26.02	1870	487	3
F	223	0	0	7870	0	6
	Check	911	100%			

After computing the composite risk for each department, risk ranking is done based on the department with the higher composite risk. Composite risk will be zero for the reference department in which the risk ranking will be the least. In the current example, composite risk is highest for department A. This implies that more amount of money is required to control risk and initiate risk control measures in department A. Amount of money allotted for safety is distributed among the department according to the risk ranking. Computations of risk rankings for other departments are shown in Table 1.4.

The goal is to reduce the potential losses within the plant while identifying the crucial department that is responsible for higher risk. This method also helps safety executives to pay attention to those departments that are crucial. Morgan's method is one of the best employed tools for such problems, as seen in the literature and possibly the easiest method to attempt financing risks (David Brown and William Dunn, 2007).

1.11 Defeating Accident Process

Different steps involved in an accident include initiation, propagation, and termination. Initiation is the event that starts the accident. This should be reduced to avoid a large accident. The procedures to control the initiation of the events are: grounding, inerting, maintenance, improved design and training to reduce human error. Propagation is the event that expands the accidents. These events should be curtailed effectively. Some of the procedures to control the propagation include emergency material transfer, fewer inventories of chemicals, use of nonflammable construction materials, installation of emergency and shutdown installation valves. Termination is the event that stops the accident. This should be increased to have a better

Table 1.5　Fatality statistics for nonindustrial activities (Lees, 1996)

Activities	FAR (deaths/10^8 h)
Staying at home	3
Traveling by car	57
Traveling by cycle	96
Traveling by air	240
Traveling by motor cycle	660
Rock climbing	4000

control over the accident. Some of the procedures to control termination are: end of pipe control measures, fire-fighting equipment, relief system, and sprinkler systems.

1.12 Acceptable Risk

In offshore industries, risk cannot be avoided. Drilling, exploration, and production processes cannot be zero-risk zones as they have inherent factors that may lead to an unforeseen incident. Depending upon the environmental conditions prevailing, they can become an accident. It is therefore important to understand that risk is accepted in offshore industries up to a certain level. According to the regulatory norms, risk is acceptable and permissible in offshore industries. According to the United States Environmental Protection Agency, risk of one in million is acceptable for carcinogens. For noncarcinogens, acceptable risk is hazard index of lesser than one. According to the United Kingdom Health and Safety Executive, acceptable FAR is unity. It is also interesting to note that even nonindustrial activities, which are part of daily routine, have risk indicators. Fatality statistics for common nonindustrial activities are given in Table 1.5.

1.13 Risk Assessment

Risk assessment is the quantitative or qualitative value of risk, which is related to a situation and a recognized hazard. Quantitative risk assessment involves in estimating both the magnitude of potential loss and the probability of occurrence of that potential loss. Therefore, risk assessment consists of two stages, namely: (i) risk determination; and (ii) risk evaluation. Risk determination deals with numbers and hence it is a quantitative approach. Risk evaluation deals with the events and hence it is a qualitative approach.

Risk is identified by continuously observing changes in risk parameters on the existing process and therefore a continuous process. Risk estimation is done by determining the probability of occurrences and the magnitude of consequences, which is post-processing of the data identified during the former stage.

Risk evaluation consists of risk aversion and risk acceptance. Risk aversion is determined by the degree of risk reduction and risk avoidance. Risk acceptance is the establishment of risk references and risk referents. Risk references are for comparing the values and the risk referents are standards with which the risk parameters are compared. For example, let us take a specific case for risk assessment of a chemical process plant. National Academy of Sciences identified four steps in chemical risk assessment, which includes hazard identification, dose–response assessment, exposure assessment, and risk characterization.

1.13.1 Hazard Identification

It includes engineering fault assessment. Basically it is used to evaluate the reliability of specific segments of a process plant, which is in operation. It determines the probabilistic results. The method employed in hazard identification is fault tree analysis.

1.13.2 Dose–Response Assessment

This involves describing the quantitative relationship between the amount of exposure and extent of toxic injury. Hazardous nature of various materials needs to be assessed before their effects are estimated. Outcome of the dose–response assessment is a linear equation relating exposure to the disease, which is obtained by the regression analysis of the dose–response data.

1.13.3 Exposure Assessment

This describes the nature and size of population exposed to the dose agent, its magnitude, and the duration of exposure. This assessment includes the analysis of toxicants in air, water, or food.

1.13.4 Risk Characterization

Risk characterization is the integration of data and the analysis. It determines whether or not the person working in the process industry and the general public in the nearby vicinity will experience effects of exposure. It includes estimating uncertainties associated with the entire process of risk assessment.

1.14 Application Issues of Risk Assessment

Risk assessment often relies on inadequate scientific information or lack of data. For example, any data related to repair may not be useful to assess newly designed equipment. It means that even though the data available is less, still all data related to that event cannot be considered as qualified data to do risk assessment. In toxicological risk assessment, the data related to use of them in animals is not relevant to predict their effects on humans. Therefore, to do risk assessment, one uses probabilistic tools for which data size is one of the main issues. Due to the limited data available in terms of occurrences of events (as the accidents are fewer) and their consequences, risk analysts use a conservative approach. They end up overestimating the risk by using statistical approach. Alternatively, one can also estimate risk on comparative scale. Conservative approach is a quantitative risk assessment, which identifies the frequency of event and its severity. After identifying the frequency and severity, risk rankings are determined to identify the critical events. Attention is paid on risk reduction or mitigation of these events instead of examining the whole process repeatedly. This is seen as one of the effective tools of risk reduction. Comparison technique is a qualitative risk assessment, which is done by conducting surveys and preparing a series of questionnaires. Based on the survey results, risk ranking is done.

1.15 Hazard Classification and Assessment

The first step in all risk assessment or Quantitative Risk Assessment (QRA) study is the hazard identification (HAZID). The purpose of HAZID is to identify all hazards associated with the planned operations or activities (Chandrasekaran et al., 2010). It provides an overview of risk, which is useful in planning further analysis of risk assessment. It provides an overview of different types of accidents that may occur in the industry with an assurance that no significant hazards are overlooked. Some of the terminologies commonly used in hazard classification and assessment are discussed next:

> *Hazard* means a chemical or physical condition that has potential to cause damage to people, property or, environment. Hazard is a scenario, which is a situation resulting in more likelihood of an incident.
> *Incident* means loss of or contamination of material or energy. All incidents do not propagate to accidents.
> *Risk* is a realization of hazard. Incident becomes an accident.

Hazard analysis is the identification of undesired events that lead to materialization of a hazard. It includes analysis of the mechanisms by which these undesired events could occur. It also includes estimation of the extent, magnitude, and likelihood of any harmful effects.

1.15.1 Hazard Identification

Hazard identification deals with the engineering failure assessment. It evaluates the reliability of specific segments of a plant in operation to determine the probabilistic results of its operational and design failure. Fault tree analysis is one of the common forms of engineering failure assessment. Hazards that are common in oil and gas industries are not identified until an accident occurs. It is therefore essential to identify the hazards if one wants to reduce risk. Some of the frequently asked questions, which lead to hazard identification are: (i) what are hazards?; (ii) what can go wrong and how?; (iii) what are the chances that they can go wrong?; and (iv) what are the consequences, if they go wrong? Answer to the first question can be obtained by doing HAZID. The answer to the question of what can go wrong and how can be obtained by doing risk assessment, which will subsequently lead to the assessment of probability of failure. Answers to questions (iii) and (iv) will actually lead to a detailed risk assessment. It is important to document all the accidents and near-miss events occurring in the offshore industries to have a wider database. It is useful in estimating the frequency of occurrence of such accidents through detailed mathematical modeling with a higher accuracy. By documenting the accidents, consequences are also identified simultaneously, which subsequently helps in risk assessment. Hazard evaluation is a combination of HAZID and risk assessment, a flowchart is given in Figure 1.7.

Hazard evaluation can be performed at any stages of operation. It can also be performed during the preliminary stages of analysis and design of the process plant. During the initial design stages, hazard evaluation is done using Failure Mode Effective Analysis (FMEA), whereas during the ongoing operation stages, it is done using Hazard and Operability Study (HAZOP). If the hazard evaluation shows low probability and minimum consequences, then the system is called *gold-plated system*. Such systems are examples of implementation of potentially unnecessary and expensive safety equipments. As can be seen from Figure 1.7, layout of hazard evaluation, the most important step in hazard evaluation is risk acceptance. It is also complex because the level of risk acceptance is subjective to each organization and hence should be predefined. Fortunately, oil industries follow international

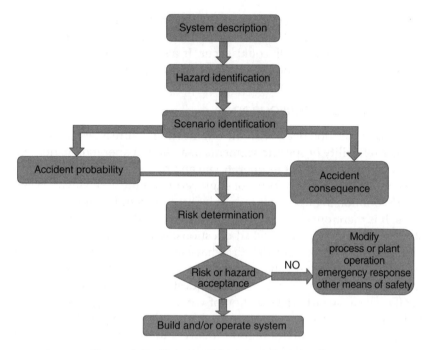

Figure 1.7 Flowchart for hazard evaluation

standards to define or determine the level of risk acceptance (OISD-169, 2011; OISD-116, 2002; OSID-144, 2005; OISD-150, 2013; OGP-2010).

Potentially unnecessary and expensive safety equipment and procedures are implemented in the system. One of the important steps in hazard evaluation is to decide on the risk acceptance criteria. It is complex as the level of risk acceptance in oil and gas industries is subjective to each organization and the process methods they adopt for exploration and production. Therefore they should be predefined even before one attempts to perform hazard evaluation. But there are also standard procedures to define or determine levels of risk acceptance.

1.15.2 Hazard Identification Methods

- *Process hazard checklists*: Refers to a list of items and possible problems in the process that must be checked periodically.
- *Hazard surveys*: Refer to the inventory of hazardous materials.
- *HAZOP*: Refers to Hazard and Operability Studies, which is carried out generally to identify the possible hazards present in any given process plant.

- *Safety review*: Refers to a less-formal type of HAZOP study. The result depends upon the experience of the person conducting the review and hence the outcome of the review can be highly subjective.
- *What-if analysis*: This is a less-formal method that applies *what-if logic* to a number of investigations. For example, the question would be *what-if the power stops?* Answers to such questions yield a list of potential consequences and solutions.
- *Human error analysis*: Refers to a method used to identify parts and procedures of a process system. It is generally applied to the process that has higher probability of human error. For example, fire alarm/buzzer system in the control panel, etc.
- *Failure Mode, Effects and Criticality Analysis* (*FMECA*): This method tabulates the list of equipments and their possible mechanical failure under working conditions. This study is capable of identifying the possible failure modes of each component present in the system and their effects of failure on the overall performance of the process system.

1.16 Hazard Identification During Operation (HAZOP)

Hazards arise due to deviations from normal process. There always exist deviations from the design intent. This is applicable to the existing and new process plants. The main purpose of HAZOP study is to identify the potential hazards and the relative operability problems that arise due to the perceived deviations. HAZOP analysis identifies all possible hazards, operational problems, recommends changes, and identifies areas that require further detailed studies. For conducting a HAZOP analysis, up-to-date Process Flow Diagram (PFDs) is required. It also requires Process and Instrumentation Diagram (P&IDs), detailed equipment specifications, details of materials and mass and energy balances. A team of experts who are experienced in a similar plant, along with the technical and safety professionals conduct HAZOP study.

1.16.1 HAZOP Objectives

HAZOP studies are carried out to identify the following:

- Any perceived deviations from intended design/operation
- Causes for those deviations
- Consequences of those perceived deviations

- Safeguards to prevent the causes and mitigate consequences of the perceived deviations
- Recommend actions in the design and operation to improve safety and operability of the plant

1.16.2 Common Application Areas of HAZOP

HAZOP is primarily used in chemical industries to estimate hazards that arise during operations; one such example can be seen in hazard studies carried out in Flixborough disaster, 1974. It is a chemical plant in the United Kingdom, which manufactures caprolactam that is required to manufacture nylon. This incident occurred due to the rupture of a temporary by-pass pipeline carrying cyclohexane at 150°C, which leaked and it set into a fire. Within few minutes, after the initiation of fire, about 20% of the plant's inventory got burnt and resulted in the spread of a vapor cloud over a diameter of about 200 m. This resulted further in an explosion of a hydrogen production plant located nearby, which showed a cascading effect of the consequences. Another similar example where HAZOP studies were applied successfully was to study the consequences of explosion at a Rocket-fuel plant located at Nevada, Las Vegas, United States, as shown in Figure 1.8. The plant was destroyed in few seconds after the initiation of explosion. The wind storm

Figure 1.8 Explosion at Rocket-fuel plant located at Nevada, Las Vegas

destroyed the roof structure and the glass. Fire was caused essentially due to the use of a welding torch in the wind-ward direction. Studies reported using HAZOP is seen to be useful in predicting the hazardous nature of the chemical release and their consequences.

1.16.3 Advantages of HAZOP

HAZOP supplements the design ideas with imaginative anticipation of deviations. These may be due to equipment malfunction or operational error. In the design of new plants, designers sometimes overlook few issues related to safety in the beginning. This may result in few errors. HAZOP highlights these errors. HAZOP is an opportunity to correct these errors before such changes become too expensive or impossible. HAZOP methodology is widely used to aid loss prevention. HAZOP is a preferred tool of risk evaluation because of few reasons: (i) easy to learn; (ii) can be easily adapted to almost all operations in the process industries; (iii) is a common method in contamination problems rather than chemical exposure or explosions; and (iv) requires no special level of academic qualification to perform HAZOP studies.

HAZOP studies examine the full description of the process thoroughly. It systematically questions every part of it to establish the perceived deviations from that of the design intent. Once identified, an assessment is made to estimate the consequences of such deviations. If considered necessary, action is taken to rectify the situation in the beginning itself. Though the method is imaginative, but it still is systematic. It is more than a checklist type of review. It encourages the team to identify possible deviations and helps to trace all of them under the operational conditions. HAZOP penetrates into greater depth of risk analysis of any process plant. HAZOP, applied on the same type of plant, repeatedly improves safety, which is quite important. Potential failures that were not noticed in the earlier studies can be easily highlighted using HAZOP.

1.17 Steps in HAZOP

Step 1: Define the design intent
Defining the design intent is the first step in a HAZOP study. Let us explain the design intent using some examples. Consider the following:

1. Suppose there is a plant in operation, which has to produce certain tons of chemical per year.

2. an automobile unit has to manufacture certain number of cars every year.
3. a plant has to process and dispose certain volume of effluent per year.
4. an offshore plant has to produce certain barrel of oil every year.

In all the cases, equipments are designed and commissioned to achieve the desired production capacity. In order to do so, each item like the equipments, pump, length of pipe work will need to be consistently functional in a particular (desired) manner. This is the design intent for that particular item and not the machinery or production capacity.

Step 2: Identify the deviations
For understanding the deviation in design intent, let us consider another example. A plant requires continuous circulation of cooling water at temperature $x°C$ and at $xxx L/h$. Cooling of the process is done by heat exchanger. For effective functioning of the plant, effective working of heat exchanger is mandatory. The design intent is the effective working of the heat exchanger. If the water supplied for circulation becomes greater than $x°C$, this would affect the production and hence this is the deviation. Note the difference between the deviation and its cause. For example, failure of pump would be a cause and not a deviation.

1.18 Backbone of HAZOP

The backbone of HAZOP studies is the keywords that are used in the study. There are two types of keywords: primary and secondary. **Primary** keyword focuses the attention on a particular aspect of the design intent or an associated process condition. **Secondary** keywords suggest possible (perceived) deviations from that of the design intent; when combined with that of the primary keywords, they intent the required meaning. As HAZOP revolves around the effective use of these keywords, it is necessary to understand their meaning and usage.

Primary keywords reflect both the process design intent and operational aspects of the plant. Examples are : FLOW, TEMPERATURE, PRESSURE, LEVEL, SEPARATE, COMPOSITION, REACT, MIX, REDUCE, ABSORB, CORRODE, ERODE, etc. These keywords sometimes may be confusing. For example, let us take the word CORRODE. One may assume that the intention is that corrosion should occur as it refers to the design intent. Most of the plants are designed with the design intent that corrosion should not occur during the life span; or if it is expected, it should not exceed a certain rate. An increased corrosion rate would result in the deviation from the design intent and therefore this word is a primary keyword. Some more primary keywords related to process are Isolate, Drain, Vent, Purge, Inspect, Maintain, Start-up, Shutdown, etc. These words are sometimes given secondary importance. For example, sometimes it is necessary to shutdown the entire plant just to re-calibrate or replace the pressure gauge in the process lines.

Secondary keywords are applied in conjunction with that of the primary to suggest the potential deviations. Examples: NO, LESS, MORE, REVERSE, ALSO, OTHER, FLUCTUATION, EARLY, etc. They convey the meaning of deviation from the design intent. For example, Flow/No indicates that there is no desired flow, which is a deviation from the design intent of FLOW. Another example could be on the operational aspect as Isolate/No. It should be noted that not all combinations of primary/secondary keywords are appropriate. For example, Temperature/No; Pressure/reverse could be considered meaningless. Results of HAZOP study are recorded in a desired format, which is termed as a HAZOP report as shown in Table 1.6.

Example problem

Let us consider an example problem of a flow line shown in Figure 1.9. FLOW/NO is applied to describe the deviation from the design intent. One of the reasons for no flow could be the blockage of the strainer S1 due to the impurities present in the dosing tank T1. Consequences that arise from the loss of dosing are incomplete separation in V1; additional causes may be cavitation in pump P1, which may result in the possible damage of the

Table 1.6 HAZOP report format

Deviations	Causes	Consequences	Safeguards	Recommended action

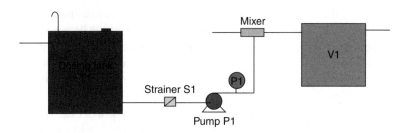

Figure 1.9 Example problem

pump, if prolonged. While recording consequences, one should be explicit. For example, instead of recording as "No dosing chemical to the mixer," it is better to add a detailed explanation along with the reason for no dosing chemical to the mixture. When assessing the consequences, one should not account for any protective systems or instruments that are already included in the design. Let us consider a case where the HAZOP team identified a cause for FLOW/NO in a system as being spurious closure of an actuated valve. It is noticed that the valve position is displayed in the central control room and also there exists an alarm in the control panel, indicating spurious closure of the valve. Even in this situation, one may think of adding the details in assessing the consequences and then recommending a few additional control measures as safeguards against the identified cause. In the example under consideration, as the spurious closure of the valve could result in the increase in pressure in the upstream line, which can lead to other cascading consequences like fire etc., it is better to add additional safeguards in spite of the presence of an alarm system in the control room. Hence, while recording HAZOP reports, one should not take the credit of the existing protective systems or instruments that are already included in the design, but to recommend additional/alternative safeguards.

Any existing protective devices, which either prevent the cause or safeguard the adverse consequences should also be recorded in the HAZOP report. Safeguards need not be restricted only to hardware; one can also recommend periodic inspection of the plants as safeguard measures. If a credible cause results in a negative consequence, it must be decided whether some action should be taken along with its priority. If it is felt that the existing protective measures are adequate, then no action need be recommended in the report.

Recording of action falls in two groups: (i) action that removes/mitigates the cause; or (ii) that eliminates the consequences. Recommended actions that address the consequences are more (the latter) as this has a direct impact on the cost control toward risk reduction. But in general, former type is preferred against the latter, but not always possible when dealing with equipment malfunction. One of the probable actions that could be recommended for the present example is to provide a strainer on the road tanker itself, which can restrict the entry of impurities to the tanker T1. However, one should be careful in such recommendations as such recommendations may result in choking the pump at the inlet section.

While recommending actions in the HAZOP report, one should not always recommend for engineered solutions such as adding additional instruments, alarms, trip-off switches, etc. It is due to the fact that any failure of mechanical systems does not resolve the actual hazard identified in the original process layout. With due regards to the reliability of such devices in operation, one should remember that their potential for spurious operation will cause unnecessary down-time. In addition, this may also result in increased operational cost in terms of maintenance, regular calibration, etc. Further complications arise if trained personnel are not appointed to operate the sophisticated protective systems; their maintenance is also equally complicated and expensive.

1.19 HAZOP Flowchart

HAZOP studies are not carried out on the whole layout of the process plant but only on the chosen segments of the plant. Usually, such segments are identified through preliminary studies such as HAZID. HAZOP procedure is discussed in the flowchart given in Figure 1.10.

1.20 Full Recording Versus Recording by Exception

HAZOP reports prepared some years ago contained partial recording of the potential deviations and the associated consequences. Some of the negative consequences were also found to be recorded as they were useful for the internal audit of the company. This method of recording reduces time and effort since they were handwritten records. Such methodology is called recording by exception. In this method, it is assumed that anything that is not included is deemed to be satisfactory. On the other hand, recent practices are to report everything in detail. Each keyword is clearly stated as applied to the system under study. Even statements like "no cause could be identified" or "no consequence arose from the cause recorded" are also seen in these statements. This is called full recording. Full recording reports verify the fact that a rigorous study has been undertaken as it is evident from the comprehensive document. This can assist in speedy assessment of safety and operability of modifications that are carried out later in the plant. With computer methods in practice, full recording has become more common these days. However, use of a few MACRO words reduces the reading time of such full records. For example, MACRO words like "no potential causes identified," "no significant negative consequences identified," "no action required," etc. can be suitable for many studies that are carried out as a part of routine maintenance.

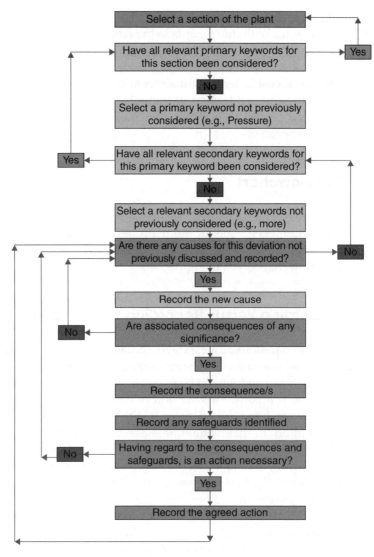

Figure 1.10   HAZOP flowchart

1.21 Pseudo Secondary Words

Pseudo secondary word is used along with the primary keyword when no appropriate secondary keyword is found suitable. For example, let us consider FLOW as one of the primary words to be used in the report.

Some combinations have credible causes, such as: FLOW/NO, FLOW/ REVERSE, etc. and a few combinations have no causes, such as FLOW/ LESS, FLOW/MORE, FLOW/OTHER, etc. So FLOW/REMAINDER can be used as a MACRO word that substitutes the meaning of a group of negation as shown in the later set. Some of the pseudo secondary words are ALL, REMAINDER, etc. After exploring all possible combinations of primary/secondary keywords, if no potential deviations could be identified, then FLOW/ALL can also be used in the report. Use of pseudo keywords improves readability as this eliminates countless repetitive entries in the report. But HAZOP report should clearly mention a list of secondary keywords in the beginning; or else, use of pseudo keywords may have ambiguous meanings.

1.22 When to Do HAZOP?

HAZOP studies are generally carried out to identify potential hazards and operability problems caused by deviations that arise from the design intent. In particular, if there are major deviations made during any recent modifications made in the process line, then the changes should be verified for their safety through HAZOP studies. As a general practice in oil and gas industries, HAZOP studies are carried out at periodic intervals of not later than 6 months. HAZOP studies should preferably be carried out as early in the design phase as possible because this influences the changes in the design if deemed fit. But unfortunately, a good HAZOP study can be carried out only on the availability of a complete design. As a compromise, HAZOP is usually carried out as a final check when the detailed design is completed. HAZOP studies may also be conducted on an existing facility to identify the modifications that should be implemented to reduce risk and operability problems. Following situations generally necessitates HAZOP studies:

- At the initial concept stage when the design and detailed drawings are available.
- When the final P&ID are available.
- During the construction and installation to ensure that valid recommendations are implemented.
- During commissioning of the plant.
- During operation of the plant to ensure that the plant emergency and operating procedures are regularly reviewed and updated as required by OSID norms.

1.22.1 Types of HAZOP

Different types of HAZOP studies are conducted depending upon the objective of the said problem. HAZOP reports should follow a set of standard procedures to make it valid under legal challenges (IEC 61882; Crawley et al., 2000; Kyraikdis, 2003). The following list explains the types along with their applicability.

> *Process HAZOP*: A technique that was originally developed to assess plants and process systems. This is quite a common type that is being practiced in oil and gas industries.
> *Human HAZOP*: A "family" of specialized HAZOPs. More focused on human errors than technical failures. Usually conducted only on violations of work permits or report of a bulk of near-miss events.
> *Procedure HAZOP*: Review of procedures or operational sequences, sometimes denoted Safe Operation Study (SAFOP). This is usually carried out while a major deviation in the process line is proposed.
> *Software HAZOP*: To identify possible errors in the development of software. This is useful to analyze the hazards that may arise from the failure of automated control systems. This is essential for all electric and electronic control systems and is often practiced in oil and gas industries.

1.23 Case Study of HAZOP: Example Problem of a Group Gathering Station

Let us consider a case study of a Group Gathering Station (GGS). Location and intrinsic details of the GGS are masked for strategic reasons but the study is actually carried out on a functional plant (Chandrasekaran, 2011c). The aim is to identify the hazards and operability problems of a GGS that has potential to cause damage to the operation, plant, personnel, and environment. The main objective is to eliminate or reduce the probability and consequences of incidents in the installation and operation of GGS. PHA-Pro7 software is used for preparing the HAZOP worksheet in the present study. Figure 1.11 shows the PFD of the GGS considered for the study. Working of the GGS is briefly explained in the following text to make the reader familiar with the process.

The well fluid emulsion, received at the limits of the GGS, is distributed into three production manifolds. From the Main Group Header, well fluid goes to the Bath Heat Treater for the first stage of separation of oil, gas, and

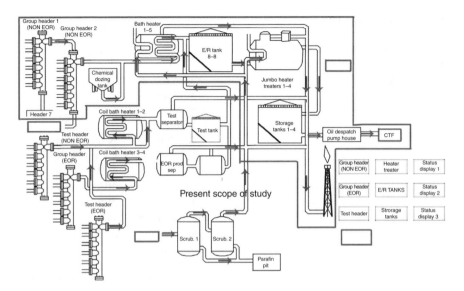

Figure 1.11 Nodes marked in the PFD of GGS

water. Separated oil is subsequently stored in the Emulsion Receipt (ER) tanks, while the associated gas is separated out and taken to the flare stack. Separated water is then drawn to the Effluent Treatment Plant (ETP) from where it is disposed after proper treatment. From the ER tanks, oil is then fed to the Jumbo Heater Treaters through the Feed pumps for refinement. In the Jumbo Heater Treaters, further separation of oil and water takes place; separated oil is then pumped to the Common Tank Form (CTF). Flow of the process line is shown the Figure 1.11.

Methodology adopted in the present study:

1. A section of the plant (NODE) on the P&ID is identified.
2. Design intent under normal operating conditions of the section is defined.
3. Deviations from the design intent or operating conditions are identified by applying a system of guide words, which are pre-defined.
4. Possible causes, related consequences, and available safeguards are identified and reported.
5. Action(s) are recommended to reduce/eliminate the deviations; focus is kept on the consequences.
6. Discussions and actions are recorded in full and detail.

HAZOP WORKSHEETS

Node 1. Group header (12″-P-102-A3A)
Deviation 1: Low or no flow
Type : pipeline

	Design conditions/ parameters:	
	1. Liquid rate:	2500 m³/day
	2. Gas flow rate	Negligible with GOR (MAX)10 V/V
	3. Pressure	10 kg/cm²
	4. Temperature	50°C
	5. Viscosity of pure oil at operating temperature	270 cp
	6. Density at operating temperature	15 API: 966 kg/m³

Causes	Consequences	Risk matrix			Safeguards	Recommendations
		S	L	RR		
1. Leak or rupture of the group header line (12″-P 102-A3A	1. Fire and environmental hazard	3	2	C	1. Fire protection systems are available	1. Pressure transmitter provided for the group header line (12″-P-102-A3A)
	2. Loss of material	2	2	C		2. Periodical hydro testing to be done for the pipeline
	3. Process upset	1	2	A		3. Periodical inspection and thickness measurement of group header line (12″-P-102-A3A) to be done

Causes	Consequences	Risk matrix			Safeguards	Recommendations
		S	L	RR		
2. Isolation valves in the inlet valves in the inlet crude oil line to the group header (12″-P-102-A3A) are stuck in closed position	1. Pressurization in the upstream section of the pipeline	1	3	C	1. Pressure gauge (PG) is available for each line from the wells	1. Pressure transmitter (PT) provided for the group header line (12″-P-102-A3A)
	2. Process upset	1	2	A	2. NRV is available for the inline to the group header (12″-P-102-A3A)	4. Periodical inspection and maintenance of the isolation valves in the inlet line to group header line (12″-P-102-A3A) to be done
					3. By-pass lines are available	
3. NRV in the inlet crude oil line to the group header (12″-P-102-A3A) is stuck in closed position	1. Pressurization in the upstream section of the pipeline	1	3	C	1. Pressure gauge (PG) is available for each line from the wells	1. Pressure transmitter (PT) provided for the group header line (12″-P-102-A3A)
	2. Process upset	1	2	A		5. Periodical inspection and maintenance of the NRV in the inlet line to the group header line (12″-P-102-A3A) to be done
4. Drain valve in the inlet crude oil line to the group header (12″-P-102-A3A) are stuck in open position or is passing	1. Fire and environmental hazard	13	2	C	1. Fire protection systems are available	1. Pressure transmitter (PT) provided for the group header line (12″-P-102-A3A)
	2. Loss of material	2	2	C		6. Periodical inspection and maintenance of the drain valve in the inlet line to the group header line (12″-P-102-A3A) to be done

(*Continued*)

Causes	Consequences	Risk matrix S L RR	Safeguards	Recommendations
5. Chocking of the inlet crude oil line to the group header (12″-P-102-A3A) due to the sludge formation	1. Pressurization in the upstream section of the pipeline 2. Process upset	1 3 C	1. Pressure gauge (PG) is available for each line from the wells	1. Pressure transmitter (PT) provided for the group header line (12″-P-102-A3A) 7. Periodical inspection and thickness measurement of the inlet line to the group header line (12″-P-102-A3A) to be done

Node 1. Group header (12″-P-102-A3A)
Deviation 2:
high flow
Type: pipeline

Design conditions/parameters:

1. Liquid rate:	2500 m³/day
2. Gas flow rate	Negligible with GOR (MAX)10 V/V
3. Pressure	10 kg/cm²
4. Temperature	50°C
5. Viscosity of pure oil at operating temperature	270 cp
6. Density at operating temperature	15 API: 966 kg/m³

Causes	Consequences	Risk matrix S L RR	Safeguards	Recommendations
1. High flow from the upstream section of this Node	1. Possibility of pressurization inside the group header (12″-P-102-A3A)	1 3 C	1. Pressure Safety Valve (PSV) is available for the Group header (12″-P-102-A3A)	1. Pressure transmitter (PT) provided for the group header line (12″-P-102-A3A)
	2. Process upset	1 2 A	2. By-pass lines are available for the header line	

Node 1. Group header (12″-P-102-A3A)
Deviation 3: Reverse or misdirected flow
Type: pipeline

Design conditions/parameters:

1. Liquid rate:	2500 m³/day
2. Gas flow rate	Negligible with GOR (MAX)10 V/V
3. Pressure	10 kg/cm²
4. Temperature	50°C
5. Viscosity of pure oil at operating temperature	270 cp
6. Density at operating temperature	15 API: 966 kg/m³

Causes	Consequences	Risk matrix S	L	RR	Safeguards	Recommendations
1. Isolation valve in the first Group header or to the testing line is stuck in open position or is passing during normal operations	1. Process upset	1	2	A		1. Pressure transmitter (PT) provided for the group header line (12″-P-102-A3A)
	2. Loss of containment	2	2	C		8. Periodical inspection and maintenance of the isolation valve in the first group header line (12″-P-102-A3A)
2. Butterfly valve connecting the two group headers is stuck in open position or is passing during normal operations	1. Process upset	1	2	A		1. Pressure transmitter (PT) provided for the group header line (12″-P-102-A3A)
	2. Loss of containment	2	2	C		9. Periodical inspection and maintenance of the Butterfly valve connecting the two group headers to be done

(*Continued*)

Node 1. Group header (12″-P-102-A3A)
Deviation 4: Low pressure
Type: pipeline

Design conditions/ parameters:

1. Liquid rate:	2500 m³/day
2. Gas flow rate	Negligible with GOR (MAX)10 V/V
3. Pressure	10 kg/cm²
4. Temperature	50°C
5. Viscosity of pure oil at operating temperature	270 cp
6. Density at operating temperature	15 API : 966 kg/m³

Causes	Consequences	Risk matrix			Safeguards	Recommendations
		S	L	RR		
1. Refer Low/ No flow deviation of this node						

Node 1. Group header (12″-P-102-A3A)
Deviation 5: High pressure
Type: pipeline

Design conditions/ parameters:

1. Liquid rate :	2500 m³/day
2. Gas flow rate	Negligible with GOR (MAX)10 V/V
3. Pressure	10 kg/cm²
4. Temperature	50°C
5. Viscosity of pure oil at operating temperature	270 cp
6. Density at operating temperature	15 API: 966 kg/m³

Causes	Consequences	Risk matrix			Safeguards	Recommendations
		S	L	RR		
1. Refer More flow deviation of this node						

Node 1. Group header (12″-P-102-A3A)
Deviation 6 : High temperature
Type: pipeline

					Design conditions/ parameters:	
					1. Liquid rate:	2500 m³/day
					2. Gas flow rate	Negligible with GOR (MAX)10 V/V
					3. Pressure	10 kg/cm²
					4. Temperature	50°C
					5. Viscosity of pure oil at operating temperature	270 cp
					6. Density at operating temperature	15 API: 966 kg/m³

Causes	Consequences	Risk matrix			Safeguards	Recommendations
		S	L	RR		
1. External Fire	1. Fire and environmental hazard	3	2	C	1. Temperature gauge (TG) is available for each of the line from the wells	10. Periodical inspection and maintenance of the fire protection system to be done
	2. Possibility of pressurization inside the group header (12″-P-102-A3A)	1	3	c	2. Pressure Safety Valve (PSV) is available for the group header (12″-P-102-A3A)	
	3. Process upset	1	3	A	3. Fire protection systems are available	

(Continued)

Node 1. Group header (12″-P-102-A3A)
Deviation 7 : Variation in composition
Type: pipeline   Design
                                                 conditions/
                                                 parameters:
                                                 1. Liquid rate :   2500 m³/day
                                                 2. Gas flow rate   Negligible with
                                                                       GOR (MAX)10
                                                                       V/V
                                                 3. Pressure   10 kg/cm²
                                                 4. Temperature   50°C
                                                 5. Viscosity of   270 cp
                                                    pure oil at
                                                    operating
                                                    temperature
                                                 6. Density at   15 API : 966 kg/m³
                                                    operating
                                                    temperature

Causes	Consequences	Risk matrix			Safeguards	Recommendations
		S	L	RR		
1. Presence of impurities in crude oil coming from wells	2. Possibility of chocking inside the group header line (12″-P-102-A3A)	1	3	C	1. Pressure gauge (PG) is available for the each line from the wells	11. Ensure arrangements for analyzing the crude oil from the wells on a regular basis are made

Risk matrix is prepared to indicate the acceptability of hazard in the GGS as per OSID norms, Figure 1.12 shows the risk matrix.

Following major conclusions are drawn from the study conducted:

• All the identified hazards of the given installation of GGS can be reduced or eliminated by implementing the suggested recommendations.
• Cost of implementation of recommendations (as calculated) influences the implementation of action significantly.
• Risk ranking of the installation is higher in Node 2 (Heater Treaters) and Node 5 (Jumbo Heater Treaters)

	Hazard severity (S)			
	No injury or health impacts (1)	Minor injury or minor health impacts (2)	Injury or moderate health impacts (3)	Death or severe injury (4)
Not expected to occur during facility life (1)	A	A	C	C
Could occur once during facility life (2)	A	C	C	N
Could occur several times during facility life (3)	C	C	N	U
Could occur on an annual basis (or more oftent) (4)	C	N	U	U

Likelihood of occurrence (L)

A - Acceptable (no risk control measures are needed)
N - Not desirable (risk control measures to be introduced within a specified time period)
C - Acceptable with control (risk control measures are in place)
U - Unacceptable

Figure 1.12 Risk matrix for the example problem (GGS)

- Recommendations of category U and N are not available in this study. Only category A and C are available. While implementing these recommendations, priority should be given to category C

While recommendations made in the above study improved the safety of operation, cost toward their implications influenced the implementation of recommended action plans (Venkata Kiran, 2011). As this being a subjective issue under the jurisdiction of the head of the operation group, implication strategies are not further discussed.

1.24 Accidents in Offshore Platforms

1.24.1 Sleipner A Platform

Consider an accident that is reported in Sleipner A platform in the North Sea. The Sleipner platform is shown in Figure 1.13. It is a condeep-type platform with concrete gravity base structure, consists of 24 cells, and has a total base area of $16\,000\,m^2$, operating at a water depth of 82 m. The platform is producing oil and gas successfully in the North Sea. Failure of the platform caused

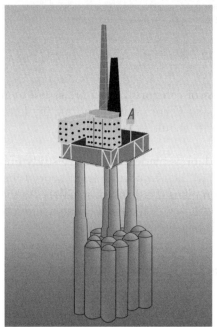

Figure 1.13   The Sleipner A platform

Figure 1.14 Thunder Horse platform

a seismic event of magnitude 3.0 on the Richter scale. The failure resulted in a total economic loss of about $700 million.

The conclusions of the investigations mentioned that the failure in a cell-wall resulted in a serious crack that propagated. Leakage was high such that the pumps were not capable to control the leakage. Wall failed as a result of the combination of a serious error in the finite element analysis and insufficient anchorage of the reinforcement in a critical zone. Shear stresses were underestimated by about 47%, leading to an insufficient design. Concrete wall thickness was reported to be inadequate.

1.24.2 Thunder Horse Platform

Another example is the accident that occurred at the Thunder Horse platform, shown in Figure 1.14. Thunder Horse production platform is located in 1920 m of water in the Mississippi Canyon Block 778/822, about 150 miles (240 km) southeast of New Orleans. Construction costs were around US $5 billion and the platform is expected to operate for about 25 years. The hull section was constructed in 2004. In July 2005, Thunder Horse was evacuated due to the threat caused by Hurricane Dennis. After the hurricane passed, the platform was inspected and assessment reports did not mention any damages to the hull of the platform. Interestingly, an incorrectly plumbed pipeline allowed

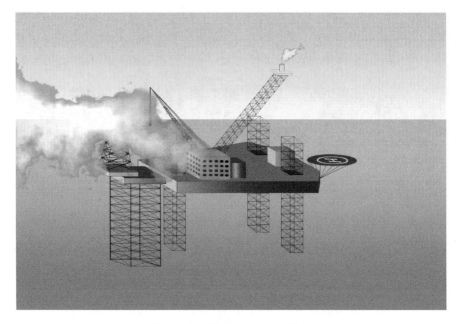

Figure 1.15　Timor Sea oil rig

water to flow freely among the several ballast tanks, which initiated the platform to tip into water. As a serious consequence of the accident, world oil prices increased because of speculation of oil shortage. The platform was subsequently rehabilitated within a fortnight after Hurricane Dennis and subsequent hits by Hurricane Katrina, 6 weeks later, did not damage the platform.

1.24.3 Timor Sea Oil Rig

Another example is the accident that occurred in Timor Sea oil rig, shown in Figure 1.15. Leaking Timor Sea oil rig caught fire on November 2, 2009. While the oil spill resulted in severe environmental damage, the cause of fire was not known immediately; personnel onboard were moved out safely without any fatal injuries.

1.24.4 Bombay High North in Offshore Mumbai

A massive platform, Bombay High North (BHN) in offshore Mumbai High field was gutted in a devastating fire on July 27, 2005. In less than 2 hours, BHN was reduced to molten metal as shown in Figure 1.16. The platform remained a beehive of activity for 24 years, which was brought to a halt due to the accident; it was later retrofitted and made functional.

Figure 1.16 Burnt out of Bombay High North platform

From the events discussed, it is important to know that the causes of failures are unknown in most of the cases. Even post-accident studies could not trace out the fundamental causes of the accident but hinted toward a set of complex reasons (Prem et al., 2010). However, from an engineering perspective, one can understand that the causes are mostly due to the oversight either in the design stage or during operation/maintenance (Valerie and Cary, 1991). As the consequences of such accidents lay serious impact on world's economy through oil pricing, it is imperative to note that risk analyses are becoming increasingly important to ensure that at least such events are not repeated (Terje and Jan Erik, 2007). It shows the importance or necessity of QRA tools (e.g., HAZOP) and their applicability to offshore platforms or any process industry in general at different stages: (i) front end engineering design stage; (ii) fabrication, construction, and commissioning stage; and (iii) operational stage etc.

1.25 Hazard Evaluation and Control

Every type of hazard is associated with some risk, which can potentially result in moderate to serious consequences. It is important to analyze the seriousness of the consequences in terms of operational, strategic, and economic perspectives. Subsequently, planning can be made to either mitigate or control the risk to an acceptable level. Hazard evaluation can be done at

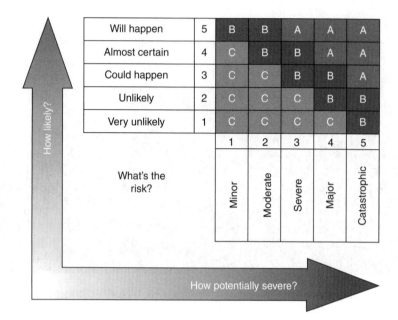

Figure 1.17   Hazard evaluation

any stage in a process plant. It can be either done during the initial design stages by conducting FMEA/FMECA or during the ongoing operation of the project through HAZOP. If the hazard evaluation shows low probability and minimum consequences, then the system is attributed as a gold-plated system, indicating that unnecessary and expensive safety equipments and procedures are implemented in the system (Skelton, 1997).

1.25.1 Hazard Evaluation

Hazards can be defined as physical or chemical conditions that have the potential to cause damage to people, property, or environment. The first step in controlling any hazard is to determine the magnitude of risk associated with it. This is often called as hazard evaluation. A simple way of evaluating hazard is assessing the total consequences associated with the hazard and the likelihood that those consequences will occur. Figure 1.17 illustrates the relationship between them.

1.25.2 Hazard Classification

Class "C" hazards pose relatively lesser risk.
Class "B" hazards pose serious risks, which means that immediate steps need to be taken to control such hazards.

Class "A" hazards are intolerable. This implies that the work should be immediately stopped until satisfactory level of hazard control is achieved.

The class into which hazards fall is the basis for deciding how to prioritize the plans for controlling them. An effective way of assessing risks and prioritizing plans for dealing with them depends on the hazard classification as shown in Figure 1.17. Other factors that influence hazard evaluation are: (i) frequency of exposure; (ii) duration of the exposure; and (iii) diagnosing additional circumstances that might affect risk like climatic conditions, etc.

1.25.3 Hazard Control

As it is difficult to eliminate hazards completely from oil and gas industries, most often attempts are made to manage hazards efficiently. The steps to manage hazards efficiently are as follows:

The first step is toward eliminating the hazard. For instance, if any damaged equipment is causing a hazard, one can think of either replacing it or by-passing it in the process line.

The second step is toward substituting hazardous materials with safer ones. This deals with the inventory control and also linked with process. For example, one can plan to replace a cleaning solution that gives off toxic fumes by a nontoxic alternative.

The third step is to isolate personnel and public from perceived hazards. A variety of steps and measures can be planned in this line to minimize hazards that can be caused to people working onboard and also to the public who live in the vicinity.

The fourth step is to adopt engineering controls to minimize risks.

The fifth step is to use administrative tools to minimize hazards. This can be done by creating more warning signals and signboards.

The sixth step is to administer protective equipments or clothing in case all the other five steps fail. This is a line of defense and therefore not the first.

This step-by-step procedure is known as Hierarchy of Hazard Controls and helps in finding the most reasonable and effective way to minimize risk of injury. In any given situation in which a hazard cannot be brought fully under control, employers are required to provide written instructions to support safe work. It is also important to ensure that workers receive sufficient level of training and supervision that is required to work safely. A Hazard Control Form will help to chalk the hazard control plan, which explains the roles and responsibilities of each team on duty to manage hazards under any unforeseen emergency.

1.25.4 Monitoring

Recommendations prescribed to control hazards need to be reviewed periodically to ensure that they are effective and appropriate. This can be a part of the ongoing regular safety inspection program. Alternatively, Joint Health and Safety Committee are formulated in many oil companies to review the control measures. Following points may be useful while reviewing the hazard controls:

- Is the hazard under control?
- Have the steps taken to manage it solved the problem?
- Are the risks associated with the hazard under control too?
- Have any new hazards been created?
- Are new hazards being controlled appropriately?
- Do workers know about the hazard?
- What has been done to control it?
- Do workers know what they need to do to work safely?
- If there is a new hazard, are workers trained properly to deal with it?
- Are there written records of all identified hazards, their risks, and the control measures taken?
- What else can be done?

Exercises 1

1. Occurrence of single or sequence of events that produce unintended loss is called ....

   Accident

2. Chemical or physical condition that has potential to cause damage to people, property, or environment is called ....

   Hazard

3. Measure of expected effects of the results of an incident is called as:
   (a) Hazard   (b) Consequence   (c) Failure   (d) Incident

   (b) Consequence

4. The relationship between the frequency and number of people suffering a given level of harm from realization of hazard is called as ....

Societal risk

5. Estimation of uncertainties associated with the entire process of risk assessment is called as ....

Risk characterization

6. ....can be a suitable tool for evaluating industrial fire risk and prioritizing units in general level of an industrial complex especially chemicals company.

Frank and Morgan risk analysis.

7. The control score for a department in an oil and gas industry is given as 156 and hazard score is 152. Calculate the percentage risk index?
(a) 24.04   (b) 26.02   (c) –26.02   (d) –24.04

(d) –24.04

8. Action taken to control or reduce risk is called ....

Risk aversion

9. In the context of a risk assessment, what do you understand by the term risk?
(a) An unsafe act or condition
(b) Something with the potential to cause injury
(c) Any work activity that can be described as dangerous
(d) The likelihood that harm from a particular hazard will occur

(d) The likelihood that harm from a particular hazard will occur

10. .... are used for representing societal risk.

FN curves

11. Prevention of hazard occurrence through proper hazard identification, assessment, and elimination is called ....

Safety

12. Define individual risk and societal risk.

Individual risk: Defined as frequency at which individuals may be expected to sustain a given level of harm from realization of hazard. It usually accounts only the risk of death. It is expressed as risk per year.

Societal risk: Defined as a relationship between the frequency and number of people suffering a given level of harm from realization of hazard. Societal risks are expressed as: FN curves, showing relationship between the cumulative frequency (*F*) and number of fatalities (*N*).

13. What is the difference between safety and risk?

Safety or loss prevention: Prevention of hazard occurrence (accidents) through proper hazard identification, assessment, and elimination.

Risk: measure of magnitude of damage along with its probability of occurrence.

14. What are the application issues of risk assessment?

Risk assessment often relies on inadequate scientific information or lack of data. For example, any data related to repair may not be useful to assess newly designed equipment. It means that even though the data available is less, still all data related to that event cannot be considered as qualified data to do risk assessment.

15. State a few golden rules of good HSE Management program.

Identifies and eliminates existing safety hazards

Safety knowledge, safety experience, technical competence, safety management support, commitment to safety

16. What do you understand by loss? What do you understand by acceptable risk? As an employee of an oil industry, how do you react to the term acceptable risk?

Loss: Severity of negative impact

Acceptable risk: Level of human and/or material injury or loss from an industrial process that is considered to be tolerable by a society or authorities in view of the social, political, and economic cost–benefit analysis.

17. Explain about safety assurance and safety assessment methods.

Safety assurance: Is the application of safety engineering practices intended to minimize the risks of operational hazards. Strategies include reactive, proactive, predictive, and iterative. Risk analysis is one of the methods.

Safety assessment: Assessed to their potential severity of impact (generally a negative impact, such as damage or loss) and to the probability of occurrence. Methods: risk assessment, hazard identification, risk characterization, etc.

18. What are goal-setting regimes and rule-based regimes?

Goal-setting regimes: Dutyholder assesses risk. Should demonstrate its understanding, controls cover management, technical, and systems issues. Keeps pace with new knowledge. Opportunity for workforce engagement.

Rule-based regimes: Legislator sets the rules. Emphasizes compliance rather than outcomes. Slow to respond. Less emphasis on continuous improvement. Less workforce involvement.

19. Explain the importance of safety in HSE management through a schematic illustration.

Importance of safety......

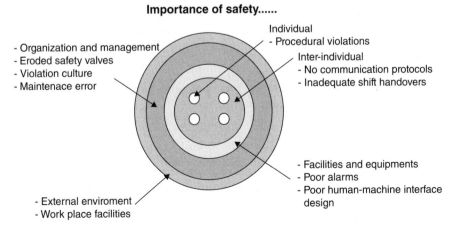

- Organization and management
- Eroded safety valves
- Violation culture
- Maintenace error

Individual
- Procedural violations
Inter-individual
- No communication protocols
- Inadequate shift handovers

- Facilities and equipments
- Poor alarms
- Poor human-machine interface design

- External enviroment
- Work place facilities

20. Calculate the risk ranking for each department?

Exposure dept	Hazard score	Control score	Property value (×10³)	Business interruption cost (×10³)	Composite score	
					Personnel	Exposure (dollars)
A	257	304	2900	1400	900	5200
B	71	239	890	1200	653	2743
C	181	180	1700	720	1610	4030
D	152	156	290	418	642	1350
E	156	142	520	890	460	1870
F	113	336	2910	3100	1860	7870

A-1, C-2, E-3, D-4, B-5, F-6.

21. Influx of fluids from the formation into the wellbore is called as
....

Well kick

22. Offshore reserve that can't economically support installation of fixed drilling and production platforms is called as ....

Marginal field

23. What are the challenges in offshore drilling?
(a) Complex operations   (b) Innovative equipments   (c) Skilled labor   (d) All of the above

(d) Complex operations, innovative equipments, skilled labor

24. Influx of fluid from the formations into wellbore is called:
(a) Dispersion   (b) Diffusion   (c) Well kick   (d) Blowout

(a) Dispersion

25. ....maintain control over potential high-pressure condition that exists in the formation.

BOP

26. What are the important factors in drilling from a safety point of view?

System design is "complete integration of all parts into the whole which should be considered in the beginning itself." Consultations are required between field development engineers, equipment manufacturers, service engineers, maintenance engineers, drilling companies, reservoir engineers, etc.

27. List different problems associated with offshore drilling operations. Also comment on the recent development of alternate drilling techniques to improve safety in operations.

   • Highly complex and technically challenging operation.
   • Uses innovative equipments and techniques.
   • Require highly special individuals to design/execute the drilling operation.

28. Three systems are commonly used as a measure of accident. What are they? Name them. Also indicate the most important common feature between them.

   • OSHA (Occupational Safety and Health Administration, US Dept of Labor)
   • Fatal Accident Rate (FAR)
   • Fatality rate or deaths per person per year
   • All three methods report number of accidents and/or fatalities for a fixed number of working hours during a specified period.

29. What are the steps taken to defeat an accident process? List different types of risk, as identified in risk analysis studies.

   Different types of risk includes strategic, financial, compliance, operations.

Defeating accident process

Steps	Desired effects	Procedure to control
Initiation (the event that starts the accident)	Diminish	Grounding, inerting, maintenance procedure, process design, training to reduce human error
Propagation (events that expand the accidents)	Diminish	Emergency material transfer, less inventory of chemicals, use non-flammable construction materials, installation of check and emergency shut down valves
Termination (events that stop the accident)	Increase	End of pipe control measures, firefighting equipment, relief system, sprinkler systems

30. What are the advantages and disadvantages of through the leg drilling?

Advantages

- Early production for improved cash flow
- Several wells in a leg can be completed and placed in production
- Drilling rig moves to a well cluster in another leg
- When wells in the 2nd leg are drilled and completed, they can be placed in production
- Continuous flow is maintained
- Time and money savings if two rigs are used
- Use a normal rig for drilling and lighter rig for completion works
- While completion rig completes the work while drilling proceeds in another leg well cluster
- Elapsed time can be reduced
- Cost savings—due to reduced on-site requirement of heavier drilling rigs

Disadvantages

- Limited to size of the completion equipment used
- Major limitation
- Number of wells that can be practically installed in a given leg

31. ....is first step in all risk assessment or QRA study.

HAZID

32. If the hazard evaluation shows low probability and minimum consequence, then the system is called as ....

Gold plated

33. ....identifies potential hazards and operability problems due to deviations.

HAZOP

34. ....is a logical, structured process that can help identify potential causes of system failure, such as causes of initiating events or failure of barrier systems.

FTA

35. .... is a most commonly used probabilistic analysis method used for hazard identification.

FTA

36. Which one of them is a primary keyword?
(a) More   (b) Reverse   (c) Erode   (d) Fluctuation

(c) Erode

37. What is a HAZARD?
(a) Where an accident is likely to happen
(b) An accident waiting to happen
(c) Something with the potential to cause
(d) The likelihood of something going wrong

(c) Something with the potential to cause

38. What are the different hazard identification methods? Explain them briefly.
- Process hazard checklists
- Hazard surveys
- HAZOP
- Safety review

39. Explain about hazard control, hazard evaluation, and hazard monitoring.

Hazard control: Sometimes hazard can be eliminated altogether, but most often measures have to be put in place to manage hazard efficiently and it also helps to be systematic. This is a step- by-step procedure which starts from the big ones, like whether to repair or upgrade the equipment and working down until you find a practical solution.

Hazard evaluation: Hazard evaluation can be performed at any stage. If the hazard evaluation shows low probability and minimum conse-quence, then the system is called gold-plated. Potentially unnecessary and expensive safety equipment and procedures are implemented in the system.

Hazard monitoring: Hazard controls need to be reviewed periodically to make sure they are still effective and appropriate. This can be part of your regular safety inspections. Talking with staff and the Joint Health and Safety Committee (if you have one) is an excellent way to start to get an idea about how well controls are working and what could be done even better. Some questions to consider when reviewing hazard controls are:

- Is the hazard under control?
- Have the steps taken to manage it solved the problem?
- Are the risks associated with the hazard under control too?
- Have any new hazards been created?

40. What is meant by hazard analysis?

- Identification of undesired events that led to materialization of a hazard
- Analysis of the mechanisms by which these undesired events could occur
- Estimation of the extent, magnitude, and likelihood of any harmful effects

41. .... is a rating corresponding to seriousness of an effect of a potential failure.

Severity

42. The objective of FMEA is on ....and not on.................

Failure prevention, and detection

43. Write short notes on HAZID and its limitations (if any).

- Deals with engineering failure assessment
- Evaluate the reliability of specific segments of a plant operation
- To determine probabilistic results of failure
- Faulty tree analysis is one such common form of engineering failure assessment
- Limitations: It is not identified until an accident occurs

44. Name one method of hazard evaluation used for mechanical and electrical systems.

FMEA

45. What do you understand by a weak link? This is required to be identified in what kind of hazard studies?

- Weak link will be the one that has the highest rank of failure
- Do a detailed analysis of the components present in the weak link
- One may also re-design to reduce the probability of failure of the components in the weak link.

This is identified while conducting FMEA

46. Name two types of FMEA.

Design FMEA, Process FMEA

47. What advantages HAZOP has when applied to a new design?

- HAZOP supplements the design ideas with imaginative anticipation of deviations. These may be due to equipment malfunction or operation error.
- In the design of new plants, designers overlook few issues related to safety in the beginning. HAZOP highlights these errors.
- HAZOP is an opportunity to correct these errors before such changes become too expensive or impossible. HAZOP methodology is widely used to aid LOSS PREVENTION.
- HAZOP is a preferred tool of risk evaluation

48. Draw a FMEA cause and effect diagram for an airbag used in passenger car.

**FMEA cause and effect diagram
Example 2 — air bag in passenger car**

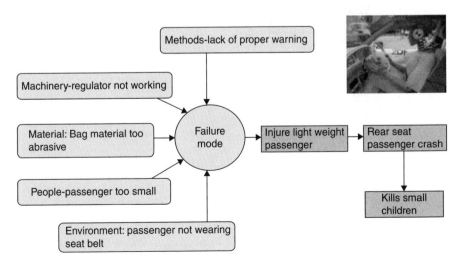

49. Explain full recording and recording by exception.

Full recording: Later practices were to report everything. Each keyword is clearly stated as applied to the system under study. Even statements like "no cause could be identified" or "no consequence arose from the cause recorded" are seen in these statements.

Recording by exception: In earlier HAZOP reports, only potential deviations with some negative consequences were recorded. Also, for handwritten records, it certainly reduces the time—both in study itself and subsequent production of HAZOP report. In this method, it is assumed that anything that is not included is deemed to be satisfactory.

50. Conduct FMEA analysis for the anti-skid braking system. The figure shows the layout plan of passenger car anti-skid braking system. Objective is to prevent locking of front wheels during heavy braking under bad road conditions. Speed sensors S1 and S2 measure the speed of two front wheels. S3 measures speed of the drive shaft. This also indicates speed of the rear wheel. Signals from three speed sensors are fed to a microcomputer. If the speed of front wheels fall significantly low, indicating application of brakes, then valves V1 and V2 are opened to reduce the braking force.

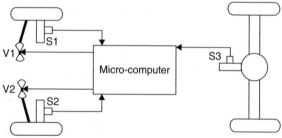

FMEA-anti-skid braking system of Car

Component	Failure mode	Failure effect	Comment
Front wheel sensor S1, S2	No output signal	Computer will assume that one wheel has stopped.	Uneven braking on front wheels
		Sends a signal to open relief valve on that wheel.	Alarm system required to switch off computer
		Results in partial loss of front wheel braking	
Front wheel valves V1, V2	Fail to open	One front wheel could lock on heavy braking	Not desired. Test facility required
	Fail to close	Partial loss of front brake	Uneven braking on front wheels Additional stop valve required?

Component	Failure mode	Failure effect	Comment
Rear wheel sensor, S3	No output signal	Microcomputer will have no reference speed from rear wheel	Alarm system required
		Will not attempt to close V1 or V2	
		Both front wheels could lock on heavy braking	
Microcomputer	No output signals to either front wheel valves	Both front wheels could lock on heavy braking	Alarm system required
	No output signal to one front wheel valve	One front wheel could lock on heavy braking	Alarm system required
	Spurious output to both front wheel valves	Total loss of front wheel braking	Alarm system required to switch off computer
	Spurious output to one front wheel valve	Partial loss of front wheel braking	Alarm system required to switch off computer

Model Paper

1. Identify major ways to prevent accidents resulting from fire and explosions.

2. Three systems are commonly used as a measure of accidents. What are they? Name them. Also indicate the most important common feature between them.

3. Define individual risk and societal risk.

4. What do you understand by acceptable risk? As an employee of an oil industry, how do you react to the term acceptable risk?

5. You are given two options to reach Station A from Station B.

   (a) You wish to drive the complete distance of 2200 km at an average speed of 45 km/h to reach Station A by road; (b) alternatively you plan to fly and reach Station B by a commercial airlines in 2½ h.

   Answer the following questions:
   1. Which travel is the safest, based on the FAR in general? Explain. Refer table for fatalities of different modes of transport.
   2. What is the fatality rate for the safest trip?
   3. Suppose you travel by car at an average speed of 60 km/h, do you think FAR will change? Will it increase or decrease? Guess the answer to this question on the basis of calculations did for the previous questions.

   Justify your answer without working out the FAR in detail.

Activity	FAR (deaths/10^8 h)
Staying at home	4
Traveling by car	57
Bicycle riding	96
Traveling by air	240
Motorcycle riding	660
Rock climbing	4000

6. An employee works in a process industry with an FAR of 4. This industry has normal working hours. As the employee gained experience in his

trade, he wishes to change his job. Another oil and gas company abroad offered him a job. The work agreement of the new company says that his working hours are only 4 hours per shift and he will have to work only for 200 days in a year.

- For your reference, see table showing the FAR for different industries
- The employee is confused as he foresees a higher risk rate in oil and gas industry compared to the current process industry where he is employed. But he expects a good financial gain.

Answer the following:

- Should the employee opt for change in his job? Being an HSE consultant, should you advise him to do so, explain the basis on which you will work out his safety in the new job.
- Suppose the employee wants to shift back to his original employer after his abroad assignment is over, should you advise him to bargain toward his working hours so that he faces the same fatality rate as that of his recent abroad assignment? If so, state briefly your advice to him.

Table: FAR for industry

Industry	FAR
Chemical industry	2
Factory work	4
Coal mining	8
sea fishing	40
Offshore oil and gas	62
Steel fabricators	70

2

Environmental Issues and Management

2.1 Primary Environmental Issues

The primary environmental issues are the huge impact caused by the oil and gas production on the shelf ecosystems and the marine biological resources. It contributes to the disturbance of life hierarchy at different levels and also significantly influences the marine ecosystem. Most importantly, it is to be noted that the biological consequences of accidental oil spills into the marine environment are irreversible.

2.1.1 Visible Consequences

Environmental pollution in the marine ecosystems creates complexities and a variety of emerging problems in the environment management. This results in an uneven distribution of marine life and its concentration in the shelf and coastal zone, which is the habitat for about 90% of the marine commercial organisms. Most of the known oil and gas fields are also located in this zone, causing serious ecological disturbances.

2.1.2 Trends in Oil and Gas Resources

The crude oil and natural gas plays a major and most important role in contributing to the total energy produced in the world. This is still increasing due

Health, Safety, and Environmental Management in Offshore and Petroleum Engineering, First Edition.
Srinivasan Chandrasekaran.
© 2016 John Wiley & Sons, Ltd. Published 2016 by John Wiley & Sons, Ltd.
Companion website: www.wiley.com/go/chandrasekaran/hse

Table 2.1 World's energy resources

Source of energy	In 1989	Optimal in 2030
Oil	33	14
Coal	24	8
Gas	18	18
Renewable sources	20	60
Nuclear power	50	0

to the high demand and increased consumption of energy. Their historical development is remarkable for its high dynamics, rapid technological progress, wide geography of exploration, and wide production activities.

2.1.3 World's Energy Resources

The world's energy resources are given in Table 2.1. It is seen from the table that there is a significant growth and relative stabilization in the recent past; decrease in oil production in large regions is also significantly noticeable (Cairns, 1992; Vinnem, 2007b). Hydrocarbon exploring fields located inland are depleted and the focus is shifted toward the shelf resources. This shift to the continental shelf is foreseen to affect the growth of marine organisms significantly (Patin, 1999). It is also a known fact that improvement in the drilling technologies led exploration possibilities to the Polar region. Advanced technology and latest equipments used for developing offshore hydrate resources pose serious threat to the environment. Many of the mechanical and chemical techniques that are used for oil exploration, production enhancement, and processing cause severe environmental issues. For example, hot water pumping and introduction of inhibitors like methanol have posed serious challenges to the marine environment. Continental shelf, which was the main arena for shipping and fishing, is now being explored for oil and gas. Prospective locations of oil and gas fields in the shelf zones often overlap with the regions of high biological productivity. Recent exploration of gas hydrates, which are highly promising, are found in the marine regions; their development is envisaged as a potential threat to the marine environment.

2.1.4 Anthropogenic Impact of Hydrosphere

Anthropogenic impact refers to assessing the state of hydrosphere and water ecosystems. It depends on many criteria such as changes in temperature regime, radioactive background, discharges of toxic effluents, inflow of nutrients, irretrievable water consumption, damage of water organisms during seismic surveys, landing of commercial species and their cultivation, destruction of the shoreline, etc. The anthropogenic impact on hydrosphere by offshore oil and gas production is given in Table 2.2 and that of the land oil and gas is given in Table 2.3.

Anthropogenic impact on marine and fresh water system causes hidden disturbances to the natural structure and function of water communities. This leads to changes in the composition and characteristics of the biotopes. Alterations in the hydrological regime and the geomorphology of water bodies are also reported in the literature. One of the serious consequences of this effect is on the fish habitat, which results in the decrease in fish population. In addition, recreational values also decrease; this may result in other ecological, economic, and socioeconomic consequences.

2.1.5 Marine Pollution

Marine pollution includes those that arise from offshore oil and gas production and marine oil transportation. Pollutants quickly spread over a large distance from the source in the open sea unlike in case of soil where it is fixed to a specific location. Most undesirable aspect of marine pollution is that when it happens it is too late to take any corrective measure.

2.1.6 Marine Pollutants

Marine pollutants can be grouped in the increasing order of hazard as follows:

- *Group 1* refers to those substances causing mechanical impacts that damage respiratory organs, digestive systems, etc. For example, suspensions, films, solid wastes, etc.
- *Group 2* refers to those substances provoking eutrophic effects and results in the rapid growth of phytoplankton. This causes disturbance of eco-structure and affects various functions of water ecosystems. For example, mineral compounds, organic substances, etc.
- *Group 3* includes those substances that have saprogenic properties (sewage with high content of easily decomposing organic matter), which causes oxygen deficiency.
- *Group 4* includes substances causing toxic effects, which causes irreversible damage to physiological process and functions of reproduction. For example, heavy metals, chlorinated hydrocarbons, etc.
- *Group 5* includes those substances that cause carcinogenic, mutagenic, and teratogenic effects. For example, benzo(a)pyrene and other polyclinic aromatic compounds, biphenyls, etc.

Table 2.2 Anthropogenic impact on hydrosphere by offshore oil and gas production

Activity	Sanitary-hygienic			Ecological			Fisheries		
	L	R	G	G	L	R	L	R	G
Liquid and solid waste discharge	—	—	—	—	Weak	Weak	Weak	Weak	—
Subsea pipelines emplacement (causing chemical pollution)	—	—	—	—	Considerable	Uncertain	Considerable	high	—
Offshore structure abandonment	—	—	—	—	Weak	Weak	Considerable	Considerable	—
Accidents (causing chemical pollution)	Considerable	Weak	—	—	Very high	Weak	Very high	Weak	—

G—global; L—local; R—regional.

Table 2.3 Anthropogenic impact on hydrosphere on land oil and gas production

Activity	Sanitary-hygienic			Ecological			Fisheries		
	L	R	G	L	R	G	L	R	G
Oil pollution	Considerable	—	—	Considerable	Weak	—	Considerable	Weak	—
Subsea pipelines emplacement (causing chemical pollution)	Very high	—	—	Very high	—	—	Very high	Weak	—

G—global; L—local; R—regional.

Table 2.4 Scale of marine pollution components

Type of impact	Scale of distribution	Sanitary	Eco-fisheries	Sources
Oil slicks, tar balls	Local	Considerable	Considerable	Oil production and transportation
Suspended solids	Local, regional	Considerable	Considerable	Bottom dredging, offshore structure emplacement, drilling
Oil hydrocarbons: crude oil and oil products	Local, regional, global	Considerable	Considerable	Oil production, storage, marine transportation
Hydrocarbons of methane series	Local, regional	Weak	Considerable	Natural gas production

Table 2.5 Level of contaminants in µg/l in surface waters

Ecological zone	Oil hydro carbons	Chlorinated hydrocarbons	Metals		
			Mercury	Lead	Cadmium
South zone	$<10^{-1}$ to 1	$<10^{-4}$ to 10^{-3}	10^{-4} to 10^{-2}	10^{-3} to 10^{-2}	10^{-4} to 10^{-2}
Ocean pelagic area southern part	$<10^{-1}$ to 1	$<10^{-3}$ to 10^{-2}	10^{-4} to 10^{-2}	10^{-3} to 10^{-2}	10^{-4} to 10^{-1}
Enclosed sea open waters	<1 to 10^{-2}	$<10^{-3}$ to 10^{-1}	$<10^{-3}$ to 10^{-2}	10^{-3} to 10^{-1}	10^{-3} to 10^{-1}
Coastal zones	10 to 10^2	10^{-3} to 1	10^{-3} to 10^{-1}	10^{-2} to 1	$<10^{-2}$ to 10^{-1}

Various components that are responsible for marine pollution and their scale of distribution is given in Table 2.4.

2.1.7 Consequence of Marine Pollutants

There are different factors that contribute to the estimate of consequences of marine pollutants, namely: (i) hazardous properties of the pollutants; (ii) volume of their input into ocean; (iii) scale of distribution; (iv) pattern of their behavior in ecosystems; and (v) stability of their composition. Worldwide contaminants of marine pollutants are given in Table 2.5. The anthropogenic impact on ocean environment causes cumulative impact on oil and gas production facilities. It can be seen from the table that their effects are mostly in the local level. Marine pollution is the leading factor for anthropogenic impact on marine ecosystems. Offshore activities contribute about 2–5% of the overall pollution in ocean environment. Also anthropogenic impact increases the concentration on marine coastal areas and shelf zones.

2.2 Impact of Oil and Gas Industries on Marine Environment

To understand the impact caused by exploration and production of hydrocarbons, it is necessary to revisit various stages of oil and gas exploration and production. There are four stages in the oil and gas development:

1. *Geological and geographical survey* is a vital stage to estimate the potential of oil well for its commercial viability.
2. *Exploration* is an important stage for identifying the rig placement, exploratory drilling, plugging the well, killing of production wells, etc.
3. *Development and production* is one of the main stages which includes platform commissioning, pipeline laying, production drilling, pipeline maintenance, etc.
4. *Decommissioning* is the final stage of the oil and gas production. This includes removal of platform, well plugging, etc. when the well is drained.

Environmental impact on each stage of oil and gas development is given in Table 2.6, while oil discharges in North sea is given in Table 2.7.

A typical drilling fluid handling system is shown in Figure 2.1. Various complexities involved in the drilling operation results in high probability of pollution to marine environment. It is important to note that while every care is taken in designing an efficient drilling system, consequences arise in various stages that cause serious impact to marine environment. These are accidental and can neither be predicted nor avoided. So, the whole effort is toward reducing the consequences of such events, which are highly unexpected during drilling operations.

2.2.1 Drilling Operations and Consequences

Periodic discharge of drilling mud from a single well is about 15–30 lb, while the mud cuttings contain dry mass of about 200–1000 tons from a single well; in case of multiple wells, this is still more. Waste water discharge is about 1500 tons per day from a single production platform. Volumes of discharge in ocean in different parts of the world are given in Table 2.8.

Table 2.6 Impact on each stage of oil production

Stage	Activities	Nature of impact
Geological and geographical survey	Seismic surveys	Interference with fisheries, impact on water organisms
	Test drilling	Sediment re-suspension, increase in turbidity
Exploration	Rig placement, exploratory drilling	Discharge of pollution, interference with fisheries
Development and production	Platform placement, pipeline laying	Physical disturbances
	Drilling of production well	Operational discharges, accident spillage, physical disturbances
	Support vessel traffic	Operational emissions, discharges, disrupting marine birds
Decommissioning	Platform removal, plugging of well	Operational discharges, residual remains of the platform, impact on organisms when explosives are used

Table 2.7 Oil discharge in North Sea

Description	Oil discharge in tons per year					
	1984	1985	1986	1988	1989	1990
Drilling cuttings	23 000	26 000	20 000	22 000	16 000	14 000
Diesel-based drilling	2100	—	—	—	—	—
Drilling discharge	2000	4000	4000	6000	4000	6000
Accident spills	1000	1000	5000	4000	1000	2000

2.2.2 Main Constituents of Oil-Based Drilling Fluid

To understand the consequences caused by drilling discharge in the open sea, it is necessary to know the main constituents of oil-based drilling fluid, which are as follows:

Barite: 409 tons (61%)
    This is one of the main constituents of the drilling mud and is capable of changing the texture and erosion properties of surface sediments near the offshore drilling sites.

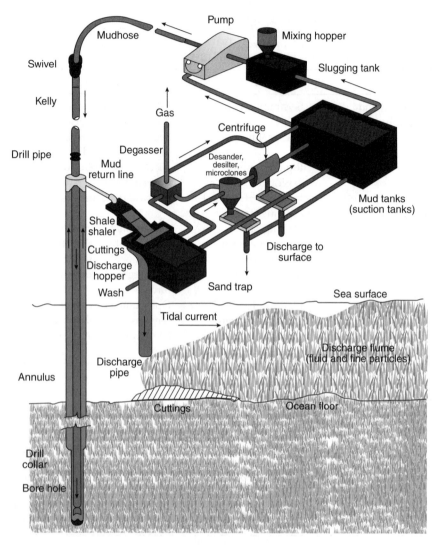

Figure 2.1 Drilling fluid handling system

Table 2.8 Volume of discharge in ocean in different parts of the world

Country	Volume of discharge (m³/day)
US, GoM	550 000
Offshore California	14 650
Cook Inlet, Alaska	22 065
North Sea	512 000
Australia	100 000

Base oil: 210 tons (31%)
Base oil in use today is formulated with diesel, mineral oil, or low-toxicity linear olefins and paraffin. The olefins and paraffin are often referred to as synthetics, although some are derived from distillation of crude oil and some are chemically synthesized from smaller molecules.

Calcium chloride: 22 tons (3.35%)
Calcium chloride is a common soluble salt used in drilling and toward the completion stages. This will be in powder, pellet, or granular form. Calcium chloride is highly hydroscopic and hence appropriate protection is important.

Emulsifier: 15 tons (2.2%)
Emulsifiers may cause emulsion blockage. This will result in increased viscosity and thus impair mobility of crude oil.

Other constituents include Filtrate agent: 12 tons (1.8%), lime: 2 tons (0.25%), and viscosifier: 2 tons (0.4%).

Each component of the drilling fluid has at least one severe technological effect. Drilling discharge contains heavy metals that have severe impact on the marine environment.

2.2.3 Pollution Due to Produced Waters During Drilling

Produced water during drilling operation contains dissolved salts and organic compounds. Along with these, oil hydrocarbons, trace metals, and suspensions are also present, which makes the composition of produced water very complex. Produced water generally contains benzene, toluene, and xylenes (10–30 mg/kg in total). *Biocides* are also present, which are used to control biological activities with limited efficiency. Organic molecules and heavy metals present in produced water is one of the important sources of marine pollution. Chromatographic analysis of the discharged water at Gulf of Mexico (GoM) showed higher and relatively stable levels of phenol and its alkylated homologues in the drilling discharges. Even radioactive elements like radium-226 and radium-228 are seen in produced waters. Radioactive elements, though of low level, remain the focus of marine pollution. During contact with sea water, these radionuclides interact with sulfates and precipitate to form a radioactive scale. They increase radioactive risk in the local and regional areas of produced waters. This affects the marine life significantly.

2.3 Drilling Accidents

Drilling accidents occur due to unexpected blowouts of liquid and hydrocarbons from the well, which results in large oil spill. One of the world's largest oil spills occurred in 1979 near the shore of Mexico after the blow out of drilling rig Ixtoc-1. This resulted in an oil spill for about 10 months, the quantity of oil spill ranged from 2500 to 6000 tons. Drilling accidents can be classified into two types, namely: (i) case leading to a catastrophic situation involving intense and prolonged hydrocarbon gushing; and (ii) case leading to routine hydrocarbon spills and blowouts that occur during normal drilling operations. Although the latter is not reported frequently, but it is responsible for causing serious pollution to marine environment.

2.3.1 Underwater Storage Reservoirs

Underwater reservoirs are used to store liquid hydrocarbons in large volumes. They are used when tankers are deployed for oil transportation instead of pipelines. Due to limitation of storage space on the topside of drilling and production platforms, underground reservoirs are commonly used. Unfortunately, risks arising from the damage of such reservoirs caused by collision of vessels, tug boats are also quite high. Such accidents usually occur during the tanker loading process or due to severe weather conditions. On damage, they become a concentrated source of marine pollution, toxicating with the methanol of high concentration. Though containing spread of such pollution is comparatively simple, their increased concentration and severe consequences to marine life in the closer vicinity is a subject matter of concern.

2.4 Pipelines

Pipelines are used to transport the explored crude oil to the shore for further processing. Owing to the increased complexities that arise from failure of pipelines, oil industries are keen in investing modern regasification plants as a part of the production facility itself. Extensive length of pipelines and inaccessibility for periodic inspection are seen as main reasons for the potential source of failure. Other factors that cause environmental risk during offshore developments due to pipelines are material defect, corrosion, tectonic movements, encountering ship anchors, and bottom vessels. Pipeline can cause a small- to long-term leakage, which can remain as a potential threat over a period of time. Intensity and scale of toxic impacts that are released on failure of pipelines vary depending upon the combination of the above factors.

2.5 Impact on Marine Pollution

Large and multiscale activity of offshore oil and gas industry imposes a serious impact on marine environment. This is a major cause of concern among the environmentalists. Impacts caused by marine pollution are chemical, physical, and biological in nature. Physical hazards can arise during the conduct of marine surveys. Seismic signals generated during marine surveys are hazardous to the marine fauna while the explosive activities of abandoned platforms result in mass migration of commercial fish. Chemical pollution is one of the major and most important impacts on marine pollution. Large offshore accidents that resulted in oil spills in the past has lead to serious ecological consequences. Moreover, fate of unused oil platforms and underwater pipelines cause serious threat to marine ecology, which is one of the passive consequences that arise from the offshore drilling activities.

2.6 Oil Hydrocarbons: Composition and Consequences

Abundant evidence seen from the published reports demonstrate the global distribution of oil contamination that primarily originated from offshore platforms. Concerns about the scale and consequences of oil pollution are increasing over the last few decades. Oil input is undoubtedly seen as one of the serious threats to marine environment. Crude oil that contains hydrocarbons with a few hybrid compositions such as paraffin-naphthenic, naphthene-aromatic, etc. are potential sources of marine pollution. Their behavior and biological impact on ecosystems are generally governed by the physical and physiochemical properties such as specific gravity, volatility, and water solubility.

2.6.1 Crude Oil

A wide difference in the properties of oil components leads to physical fractionalizing of crude oil in the ocean environment. This makes the oil to be present in different physical states: (i) as surface films; (ii) in the dissolved forms; (iii) as emulsion (oil-in-water, water-in-oil); (iv) in the suspended forms; and (v) as oil aggregates that float on the surface. Their fractions are absorbed by the suspended particles; solid and viscous components get deposited at the sea bottom and other compounds get accumulated in the water organisms.

2.7 Detection of Oil Content in Marine Pollution

One of the main problems of detecting the presence of oil content in marine pollution is the existence of hydrocarbons similar to that produced by the marine living organisms. Detection becomes even more difficult when the oil presence is low but has high background concentrations. The complex process of oil transformation starts developing from the very moment oil comes in contact with sea water. However, progression, duration, and results of these transformations depend on properties and composition of the crude oil. Spread of oil on free surface occurs under the influence of gravitational forces.

2.8 Oil Spill: Physical Review

Oil spill undergoes various stages in which each stage pollutes the marine environment significantly. Various stages include the following: (i) physical transport; (ii) microbial degradation; (iii) aggregation; and (iv) self-purification. Within few minutes of the spill of oil, it can disperse over a circumference along the free surface. It forms a thin slick of about 10 mm thick, which becomes thinner as it continues to spread further. The area of spread of oil spill can even extend to a few square kilometers. During first few days after the spill, considerable part of the spillage gets transformed into the gaseous phase. While the slick gradually loses water-soluble hydrocarbons, the remaining fraction, being viscous, reduces slick spreading. Most of the oil components like aliphatic and aromatic hydrocarbons are water soluble to a certain degree due to their lower molecular weight. Hydrodynamic and physiochemical conditions influence the rate of dissolution of oil in surface waters. Chemical transformations of oil on water surface takes place as early as within a day of the oil spill, which results in oxidative nature of reaction. This involves petrochemical reactions under the influence of ultraviolet waves of solar spectrum. Some traces of vanadium and compounds of sulfur catalyze the oxidation process. The final products of oxidation have increased water solubility and toxicity such as hydroperoxides, phenols, carboxylic acids, ketones, aldehydes, and others.

2.8.1 Environmental Impact of Oil Spill

Available data on sea water levels of oil hydrocarbons vary in different regions. Factors influencing them arise from the complexities of their bio-geo-chemical behavior. Reports show that the tendencies of oil levels tend to increase from ocean pelagic region to the enclosed sea, coastal waters, and estuaries. Marine pollution studies also identified the maximum contamination of euphotic layer, patchy distribution of contaminants, localization in upper microlayer, deposition in bottom sediments, increased levels in contact zones and overlapping fields of maximum pollution through their recent reports.

2.9 Oil: A Multicomponent Toxicant

The eco-toxicological characteristics of oil is of extreme complexity and causes variability in its composition. Oil is an important toxicant due to its integrated nature. It affects every vital function such as process, mechanism, and system of living organisms. Oil hydrocarbons with complex molecules are more toxic than that of simpler molecules and straight chain of carbon atoms. Increasing molecular weight of the components increases their toxicity. Biomarker methodology is an important element of marine monitoring, which provides data for assessing the cumulative biological effects under the chronic oil contamination of sea water.

2.9.1 Oil Spill

Oil hydrocarbons are continuously released in marine environment due to natural oil seepage from sea floor. Global distribution of oil hydrocarbons in World Ocean is characterized by increasing concentration from pelagic areas to coastal waters. From the chemical point of view, oil is a complex mixture of many organic substances, which are dominated by hydrocarbons. When they come in contact with the marine environment, they are easily separated into fractions. These separated fractions result in formation of surface slicks, dissolved and suspended substances, emulsions, solid, and viscous components. Migration of oil in biological perspective is a complex and interconnected process. They include physical transport, dissolving and emulsification, oxidation and decomposition, sedimentation and microbial degradation.

2.10 Chemicals and Wastes from Offshore Oil Industry

2.10.1 Drilling Discharges

Drilling mud is hazardous due to their persistence in marine environment. After 6 months of discharge of oil-based drilling waste, it is found that they biodegrade only by 5% (Ostgaard and Jensen, 1983). Drilling waste such as fatty acids lose their organic fraction due to microbial and physiochemical decomposition. Water-based drilling mud is generally disposed overboard, which adds more intensity to the marine pollution. Drill cuttings, which are pieces of rock crushed by the drill bit are brought to surface; they generally do not pose any special threat, but they increase the turbidity and smothering effect of benthic organisms. Oil-based mud in particular contain a wide array of organic and inorganic traces that are hazardous in nature. Discharge of drilling cuttings in large volume imposes eco-toxicological disturbances in the areas of offshore production. Oil and the oil products that are present in the drilling cuttings are the main toxic agents. Permissible limit of drilling cuttings discharge cannot exceed 100 g/kg; in reality this concentration is exceeded by about 100 times. Drilling waste that is discharged into the marine environment disperses in the solid phase. This contains clay minerals, barite, and crushed rock. Large and heavy particles are rapidly sedimented, while the small fractions gradually spread over larger distances.

Produced water is one of the forms of discharge that is evacuated from offshore platforms whose volume is significantly high. They include solutions of mineral salts, organic acids, heavy metals, and suspended particles. Produced waters, when combined with the injection water, that are used for oil recovery, cause more complications due to their mixed chemical composition.

2.11 Control of Oil Spill

The LC_{50} values (96-hour exposure) of the majority of drilling fluids vary from 10 to 15 g/kg showing lethal substances of high toxicity. Drilling fluid contains three main groups of toxicity. Group 1 refers to low-toxic substances such as bentonite, barite, and lignosulfonates. Group 2 refers to high-toxic compounds such as biocides, corrosion inhibitors, and descalers; they are seen in small proportions. Group 3 refers to the medium-toxic compounds such as lubricating oil, emulsifiers, thinners, and solvents, which are seen in large percentage. Oil spill can be controlled by many mechanical, chemical, and biological methods; mechanical methods are generally preferred due to higher efficiency. One of the most common mechanical methods of controlling oil spill is by deploying floating booms. Oil slick spreading can be controlled using the booms and oil is then collected from oil collectors. Special ships having floating separating units are used for this purpose. Usually mechanical means are supplemented by chemical spill-control methods as well.

2.12 Environmental Management Issues

Environmental issues arise from the oil and gas development activities, which is the current focus of scientific community; it also draws public attention all over the world. Environmental management policies are framed by the local and global regulatory authorities, which takes into account the factors of current and future interest. These factors include: (i) possibilities of alternative sources of energy; (ii) natural conditions; (iii) ecological factors; and (iv) techno-economic factors.

2.12.1 Environmental Protection: Principles Applied to Oil and Gas Activities

Various factors contribute to the implementation of policies and regulations that are regulatory to control environmental pollution that arise from oil and gas exploration. They are listed as follows:

- Acknowledgment of socioeconomic stipulation: Many countries are framing policies in cooperation with oil producers, fishermen, and environmentalists to achieve mutual understanding across their respective domains.
- Expediency of developing offshore natural resources.
- Using an ecocentric approach in contrast with anthropocentric approach: This is an alternative approach, which ensures stability of natural ecosystems. It supports conditions for self-renewal of biological resources.
- Environmental protection policies are governed by regional aspects accounting for specific features of different marine basins, in terms of diverse climate, social, economic, and other characteristics.

Guidelines framed by the Joint Group of Experts of Scientific Aspects of Marine Pollution are generally followed (GESAMP, 1991). These guidelines indicate three main blocks such as planning, assessment, and regulation. They include the regulatory measures for discharging drilling waste into sea. The most important guideline, which is implemented with strict compliance is that the discharges into sea require proper authorization. Concentration of oil and oil products determined using standard tests should not exceed the established standards. LC_{50} values for discharge samples established using 96-hour Mysid toxicity test should not exceed 30 g/kg (Cano and Dorn, 1996).

2.12.2 Environmental Management: Standards and Requirements

The standards that govern implementation of environmental management policies include the following:

- Content of mercury and cadmium in barite base of drilling fluid is restricted.
- No discharge of drilling waste is allowed in waters within 3 miles from the shore.
- No discharge of diesel oil is allowed.
- No discharge of free oil, based on static sheen test is allowed. Tests should conform to LC_{50} values on the basis of Mysid toxicity test.
- Average oil concentration should not be more than 7 mg/l for a monthly oil content and 13 mg/l for an average daily oil concentration.
- Discharge is to be measured within 4 miles from the shore to ascertain its toxicity compliancy.

2.13 Ecological Monitoring

Ecological monitoring is a system that collects information about the changes in natural parameters that occurred due to the oil pollution in open sea. Since contents of oil pollution cannot be measured directly due to its complex composition, its effects are measured in terms of its consequences on marine organisms; this is an indirect method of monitoring. Ecological monitoring is one of the basic methods to control and manage activities related to marine pollution. Biological monitoring is based on the technique of measuring molecular and cellular effects under low levels of impact that is not capable by chemical analysis. Ecological monitoring in offshore oil production is done at the local level. Results of the monitoring should be strictly in compliance with the established standards.

2.13.1 Ecological Monitoring Stages

Ecological monitoring is done in different stages. At the first stage, possible potential hazards that arise from the impact sources are identified. At the second stage, regular observations of marine biota are made to qualitatively assess the responses in biological organisms. The cause–effect relationship between the biological effects and impact factors are then studied. Subsequently, total impact on the marine environment and biota, including the impact on commercial species and biological resources are assessed. At the final stage, correcting measures are recommended for checking the marine pollution along with the suitable preventive measures, if any.

2.14 Atmospheric Pollution

2.14.1 Release and Dispersion Models

Release model identifies the type of release of material. This is used to assess release rate of the toxic material into the atmosphere. It also includes estimation of the down-wind concentration of the released material. *Dispersion model* describes how vapors are transported down-wind of a release. Three different kinds of vapor cloud behavior and the corresponding release time models are considered, which are given in Table 2.9.

Table 2.9 Vapor cloud behavior

Vapor cloud behavior	Release-time mode
Neutrally buoyant gas	Instantaneous (puff)
Positively buoyant gas	Continuous (plumes)
Dense buoyant gas	Time varying continuous

2.14.2 Continuous Release and Instantaneous Release (Plume and Puff Models)

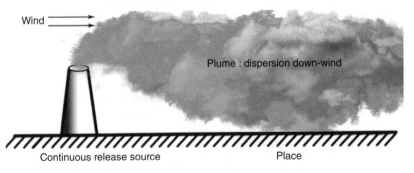

Figure 2.2 Continuous release (plume)

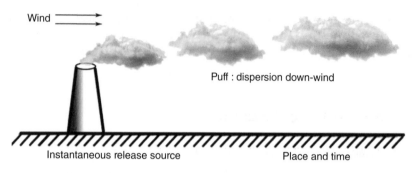

Wind

Puff : dispersion down-wind

Instantaneous release source   Place and time

Figure 2.3   Instantaneous release (puff)

2.14.3 Factors Affecting Dispersion

Parameters that affect dispersion are wind speed, terrain effects, atmospheric stability, height of release above ground, and the initial momentum of the released material.

(a) *Wind speed*
Dispersed or emitted gas initially gets diluted with the passage of air; emitted gas is then carried forward faster. In this process it also gets diluted due to addition of large quantity of air. Wind speed and the direction can be obtained from the wind rose diagram for a particular geographic location. Wind rose diagrams are plotted for specific regions, which shows wind speed, direction, and relative frequency at that region. It comprises of 16 angular wedges, each representing an arc of 22.5° segments. It contains eight colored segments in which each color represents specific wind speed when blowing from a specific direction. The overall radius of each wedge represents the percentage of time wind came from that direction during the period of interest. Wind speed and the direction of wind influence the release significantly.

The near-neutral and stable air condition for wind profile is given by:

$$U_Z = U_{10} \left(\frac{Z}{10} \right)^p \tag{2.1}$$

where, p is the power coefficient ($p = 0.4$, 0.28, and 0.16 for urban, suburban, and rural areas, respectively), U_{10} is wind speed at 10 m elevation and Z is the elevation (in m).

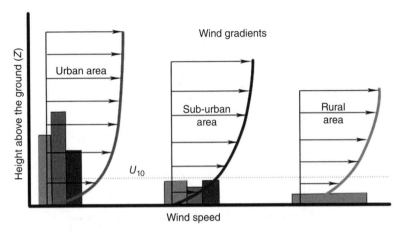

Figure 2.4 Wind profile for different areas

(b) *Terrain effects*
Ground conditions and terrain effects also influence the mechanical mixing at the surface, while height affects the wind profile. Physical interferences caused by the presence of trees and buildings increases the mixing but open sources like lakes and open ground decreases the mixing. The wind profile at different areas is given in Figure 2.4.

(c) *Atmospheric stability*
It is defined by the atmospheric vertical temperature gradient. During the day time, air temperature decreases rapidly with the increase in height; this will encourage the vertical lift motion. The lapse rate is given by:

$$\Gamma = -\left(\frac{dT}{dZ}\right) \cong 1°C / 100\,m \tag{2.2}$$

where, dT is the temperature differential and dZ is variation in height.

Atmospheric stability is classified into three groups: unstable, neutral, and stable. Under unstable atmospheric conditions, sun heats the ground faster than the heat that can be removed so that air temperature near the ground is higher than that at higher elevations. Under neutral atmospheric conditions, air above the ground warms and the wind speed is higher. This reduces the effect of solar input. Under stable atmospheric conditions, sun cannot heat the ground as fast as the ground cools. As a result of which temperature at the ground will be comparatively lower. Figure 2.5 shows the air temperature as a function of altitude for day and night conditions. Plots consider the adiabatic temperature gradient for humid air as 0.5°C/100 m.

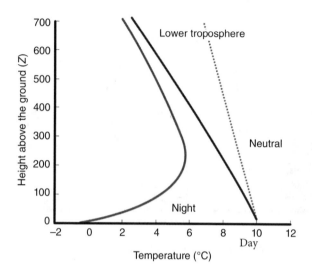

Figure 2.5　Air temperature as a function of altitude

Table 2.10　Pasquill Stability classes for day and night conditions

Surface wind speed (m/s)	Day, incoming solar radiation			Night, cloud cover thickly overcast		Anytime heavy overcast
	Strong	Moderate	Slight	>1/2 low clouds	<3/8 clouds	
<2	A	A–B	B	F	F	D
2–3	A–B	B	C	E	F	D
3–5	B	B–C	D	D	E	D
5–6	C	C–D	D	D	D	D
>6	C	D	D	D	D	D

The atmospheric stability classes are also classified according to the Pasquill Stability of atmosphere. It is classified into six categories: A, B, C, D, E, and F. A is extremely unstable condition with very low wind speed. B is moderately unstable condition. Atmospheric stability class C refers to slightly stable condition with an increase in wind velocity, while class D refers to a neutrally stable condition, which is generally used for overcast conditions. Class E is slightly stable condition, which is generally used for night conditions, while class F refers to a moderately stable atmospheric condition. Table 2.10 shows the Pasquill Stability classes for day and night conditions.

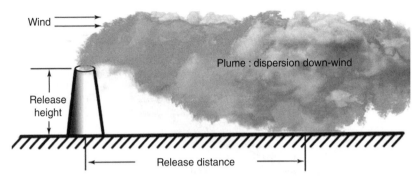

Wind

Plume : dispersion down-wind

Release height

Release distance

Figure 2.6 Release of material above ground

(d) *Height of release above ground and momentum of material*
Ground level concentration of a dispersed plume decreases with the increase of source of the release height. Figure 2.6 shows the release above the ground. Momentum of the released material depends on the effective release height and initial buoyancy. For example, momentum of high velocity jet will carry the release material with a velocity higher than that at the point of release. Gas will be initially negative buoyant and will slump toward the ground. If gas has lower density than air, it will initially be positive buoyant and will be lifted upward.

2.15 Dispersion Models for Neutrally and Positively Buoyant Gas

Neutrally and positively buoyant gas dispersion models are useful to estimate the average concentrations and predict time profile of flammable toxic gases along the down-wind direction of the release. Similar to that of liquid release models, plume and puff models are commonly used to model the vapor cloud dispersion. Plume model describes the continuous emission of materials from a steady height, H, above the ground level, which is shown in Figure 2.7. The wind blowing direction is taken along the X axis.

2.15.1 Plume Dispersion Models

The average released material or gas concentration is given by:

$$C(x,y,z) = \frac{Q}{2\Pi\sigma_x\sigma_y U}\exp\left[-\frac{1}{2}\left(\frac{y}{\sigma_y}\right)^2\right] x$$

$$\left\{\exp\left[-\frac{1}{2}\left(\frac{z-H}{\sigma_z}\right)^2\right] + \exp\left[-\frac{1}{2}\left(\frac{z+H}{\sigma_z}\right)^2\right]\right\}$$

(2.3)

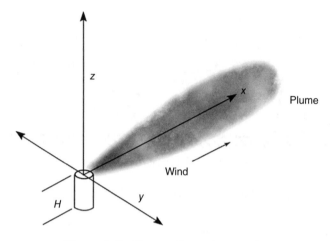

Figure 2.7 Plume dispersion model

where, $C(x, y, z)$ is the average concentration of release material (kg/m^3), H is the height of the releasing source from ground (m), (x, y, z) are distances from the source in down-wind, cross-wind, and vertical direction, respectively (m), Q is the release strength of the material (kg/s), U is the wind velocity (m/s), and (σ_y, σ_z) are dispersion coefficients in y and z directions, respectively.

Let us consider some of the cases of plume dispersion model.

Case 1: Ground level centerline concentration $(y=z=0)$, then the concentration is given by:

$$C(x,0,0) = \frac{Q}{2\Pi\sigma_z\sigma_y U}\exp\left[-\frac{1}{2}\left(\frac{H}{\sigma_z}\right)^2\right] \qquad (2.4)$$

Case 2: Ground, centerline, release height, $H=0$, then the concentration is given by:

$$C(x,0,0) = \frac{Q}{2\Pi\sigma_z\sigma_y U} \qquad (2.5)$$

In both the cases, x is implicit with the dispersion coefficients.

2.15.2 Maximum Plume Concentration

Maximum plume concentration always occurs at the release point (Brode, 1959). For releases above ground, maximum concentration occurs along the centerline (X axis) of the down-wind direction. The distance at which maximum ground level concentration occurs is given by:

$$\sigma_z = \frac{H}{\sqrt{2}} \qquad (2.6)$$

The maximum concentration is given by:

$$C_{max} = \frac{2Q}{e\Pi UH^2}\left(\frac{\sigma_z}{\sigma_y}\right) \qquad (2.7)$$

2.16 Puff Dispersion Model

The puff dispersion model describes instantaneous release of the material (consider, e.g., a sudden release of a chemical from a ruptured vessel). Consequences that arise from such a release will be the formation of large vapor cloud from the dispersed (rupture) point. In this case, classical puff model is used to describe a plume as well. The average concentration is estimated for the puff release using the following relationship:

$$C(x,y,z) = \frac{Q_{instantaneous}}{(2\Pi)^{3/2}\sigma_x\sigma_y\sigma_z}\exp\left[-\frac{1}{2}\left(\frac{x-ut}{\sigma_x}\right)^2\right]x$$

$$\exp\left[-\frac{1}{2}\left(\frac{y}{\sigma_y}\right)^2\right]+\left\{\exp\left[-\frac{1}{2}\left(\frac{z+H}{\sigma_z}\right)^2\right]+\exp\left[-\frac{1}{2}\left(\frac{z+H}{\sigma_z}\right)^2\right]\right\} \qquad (2.8)$$

Let us consider some special cases of puff modeling.

Case 1: The total integrated dose at ground level (i.e., $z=0$) is given by:

$$Dose(x,y,0) = \frac{Q_{insantaneous}}{\Pi\sigma_y\sigma_z u}\exp\left[\left(-\frac{1}{2}\left(\frac{y}{\sigma_y}\right)^2\right)-\left(-\frac{1}{2}\left(\frac{H}{\sigma_z}\right)^2\right)\right] \qquad (2.9)$$

Case 2: The concentration on ground below the puff center is given by:

$$C(x,0,0,t) = \frac{Q_{insantaneous}}{\sqrt{2}\Pi^{3/2}\sigma_z\sigma_y\sigma_y}\exp\left[-\frac{1}{2}\left(\frac{H}{\sigma_z}\right)^2\right] \qquad (2.10)$$

Case 3: Puff center on ground (i.e., $H=0$) is given by:

$$C(x,0,0,t) = \frac{Q_{\text{insantaneous}}}{\sqrt{2\Pi}^{3/2}\sigma_z\sigma_y\sigma_y} \qquad (2.11)$$

2.16.1 Maximum Puff Concentration

The maximum puff center is located at the release height and center of puff is located at x (= ut, where u is the wind velocity). On ground, maximum concentration always occurs directly below the puff center.

2.17 Isopleths

Isopleths measure the cloud boundary at a fixed concentration. It represents the lines of constant concentration as shown in Figure 2.8. Different steps to determine isopleths are as follows:

Step 1: Determine concentrations along the centerline at fixed points along the down-wind direction.

Step 2: Find the off-center distances to isopleths (y) at each point using Equation (2.12).

$$y = \sigma_y\sqrt{2\ln\left(\frac{C(x,0,0,t)}{C(x,y,0,t)}\right)} \qquad (2.12)$$

where $C(x, 0, 0, t)$ is the down-wind ground centerline concentration and $C(x, y, 0, t)$ is the isopleths concentrations at (x, y).

Step 3: Plot isopleths offset for both the directions at each point as shown in Figure 2.9.

Step 4: Connect the points as shown in Figure 2.10 to get isopleths.

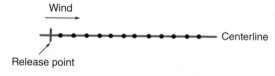

Figure 2.8 Isopleths

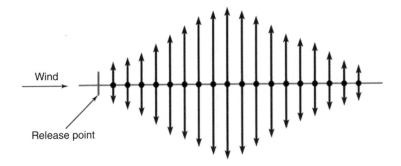

Figure 2.9   Isopleths offset

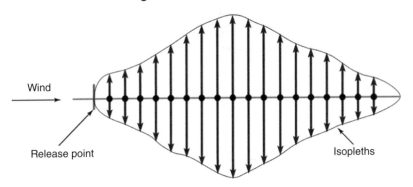

Figure 2.10   Isopleths

2.18 Estimate of Dispersion Coefficients

The dispersion coefficients are important to model the release scenarios using plume or puff models (Wiltox, 2001). They depend upon the stability class and down-wind distances. For calculating the dispersion coefficients, initially identify the Pasquill Stability class by using meteorological data such as wind speed, heat radiation, cloud cover, etc. Classification of the area as rural, urban, flat, hilly is also required. Using Figures 2.11, 2.12, and 2.13, one can estimate the dispersion coefficients for the relevant cases as applicable.

2.18.1 Estimates from Equations

Dispersion coefficients for plume model can also be estimated mathematically. Let X be the down-wind distance (in m) measured from the source of release. Dispersion coefficients can be calculated as given in Tables 2.11 and 2.12 for plume and puff models, respectively.

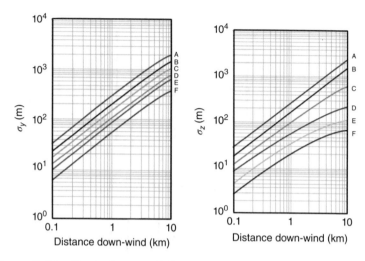

Figure 2.11   Dispersion coefficients for plume model (rural release)

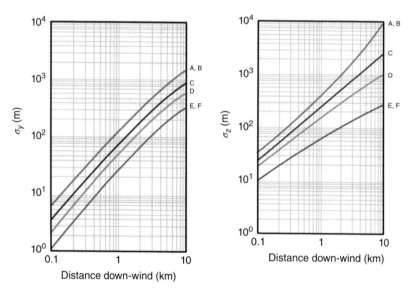

Figure 2.12   Dispersion coefficients for plume model (urban release)

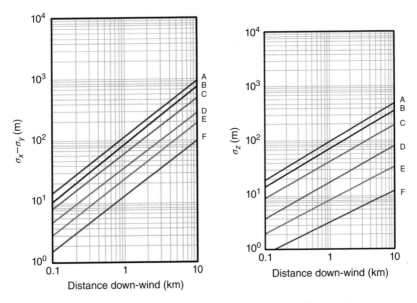

Figure 2.13 Dispersion coefficients for puff model

Table 2.11 Calculation of dispersion coefficients (plume model)

Area	Stability class	σ_y (m)	σ_z (m)
Rural conditions	A	$0.22X(1+0.0001X)^{-0.5}$	$0.20X$
	B	$0.16X(1+0.0001X)^{-0.5}$	$0.12X$
	C	$0.11X(1+0.0001X)^{-0.5}$	$0.08X(1+0.0002X)^{-0.5}$
	D	$0.08X(1+0.0001X)^{-0.5}$	$0.06X(1+0.0015X)^{-0.5}$
	E	$0.06X(1+0.0001X)^{-0.5}$	$0.03X(1+0.0003X)^{-1.0}$
	F	$0.04X(1+0.0001X)^{-0.5}$	$0.016X(1+0.0003X)^{-1.0}$
Urban conditions	A–B	$0.32X(1+0.0004X)^{-0.5}$	$0.24X(1+0.0001X)^{-0.5}$
	C	$0.22X(1+0.0004X)^{-0.5}$	$0.20X$
	D	$0.16X(1+0.0004X)^{-0.5}$	$0.14X(1+0.0003X)^{-0.5}$
	E–F	$0.11X(1+0.0004X)^{-0.5}$	$0.08X(1+0.0001X)^{-0.5}$

Table 2.12 Calculation of dispersion coefficients (puff model)

Area	Stability class	σ_x or σ_y (m)	σ_z (m)
Rural conditions	A	$0.18X^{0.92}$	$0.60X^{0.75}$
	B	$0.14X^{0.92}$	$0.53X^{0.73}$
	C	$0.10X^{0.92}$	$0.34X^{0.71}$
	D	$0.06X^{0.92}$	$0.15X^{0.70}$
	E	$0.04X^{0.92}$	$0.10X^{0.65}$
	F	$0.02X^{0.89}$	$0.05X^{0.61}$

2.19 Dense Gas Dispersion

Gases having density higher than air are termed as dense gases. Dense gases released from the source initially slump toward the ground and subsequently move upward and progress along the down-wind directions. Mixing mechanisms with air are completely different from that of the neutrally buoyant releases. Britter-McQuaid dense gas dispersion model is commonly used in such cases.

2.19.1 Britter-McQuaid Dense Gas Dispersion Model

Step 1: Characterize the initial buoyancy using the following relationship:

$$g_0 = g\left(\frac{\rho_0 - \rho_a}{\rho_a}\right) \quad (2.13)$$

where, g is acceleration due to gravity, ρ_0 and ρ_a are density of the released material and ambient air, respectively.

Step 2: Decide whether the release is instantaneous or continuous using the following relationship:

$$F = \left(\frac{uR_d}{x}\right) \quad (2.14)$$

where, u is the wind velocity, x is the distance from the release point, and R_d is the duration of the release. For $F \geq 2.5$, the release is assumed to be continuous; for $F \leq 0.6$, it is considered as instantaneous release. In case, $0.6 < F < 2.5$ is satisfied, then one can use both the approaches to find the maximum value.

Step 3: Characterize the source dimension.
For continuous release, source dimension (D_c) is given by:

$$D_c = \sqrt{\frac{q_0}{u}} \quad (2.15)$$

where, q_0 is initial plume volume flux and u is the wind speed.
For instantaneous release, source dimension (D_i) is given by:

$$D_i = V_0^{1/3} \quad (2.16)$$

where V_0 is initial volume

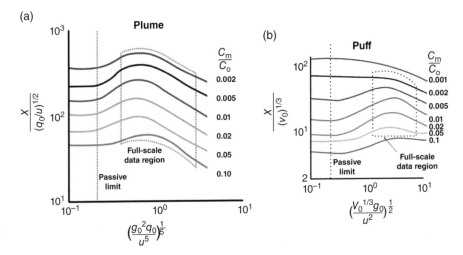

Figure 2.14 Concentration ratio: (a) for plume model; (b) for puff model

Step 4: Checking criteria
For continuous release, value is checked using the following relationship:

$$\left(\frac{g_0 q_0}{u^3 D_c}\right) \geq 0.15 \tag{2.17}$$

For instantaneous release, the check is done using the following equation:

$$\frac{\sqrt{g_0 V_0}}{u D_i} \geq 0.20 \tag{2.18}$$

If the criterion is satisfied, then the concentration ratio (C_m/C_0) is given by Figure 2.14.

Concentration ratio is also given by the relationships shown in Tables 2.13 and 2.14 for gas plume and gas puff models, respectively.

2.20 Evaluation of Toxic Effects of Dispersed Liquid and Gas

Toxicity of the dispersed liquid or gas in atmosphere is measured based on two parameters, that is, concentration of dispersion and duration of exposure (Leonelli et al., 1999). Permissible Exposure Limit (PEL) or Threshold Limit

Table 2.13 Dispersion of dense gas plumes

Concentration ratio (C_m/C_o)	Valid range for $\alpha = \log (g_o^2 q_e/u^5)^{1/5}$	$\beta = \log [x/(q_o/u)^{1/2}]$
0.1	$\alpha \le -0.55$	1.75
	$-0.55 < \alpha \le -0.14$	$0.24\alpha + 1.88$
	$-0.14 < \alpha \le 1$	$0.50\alpha + 1.78$
0.05	$\alpha \le -0.68$	1.92
	$-0.68 < \alpha \le -0.29$	$0.36\alpha + 2.16$
	$-0.29 < \alpha \le -0.18$	2.06
	$-0.18 < \alpha \le 1$	$-0.56\alpha + 1.96$
0.02	$\alpha \le -0.69$	2.08
	$-0.69 < \alpha \le -0.31$	$-0.45\alpha + 2.39$
	$-0.31 < \alpha \le -0.16$	2.25
	$-0.16 < \alpha \le 1$	$-0.54\alpha + 2.16$
0.01	$\alpha \le -0.7$	2.25
	$-0.7 < \alpha \le -0.29$	$-0.49\alpha + 2.59$
	$-0.29 < \alpha \le -0.20$	2.45
	$-0.20 < \alpha \le 1$	$-0.52\alpha + 2.35$
0.005	$\alpha \le -0.67$	2.4
	$-0.67 < \alpha \le -0.28$	$-0.59\alpha + 2.8$
	$-0.28 < \alpha \le -0.15$	2.63
	$-0.15 < \alpha \le 1$	$-0.49\alpha + 2.56$
0.002	$\alpha \le -0.69$	2.6
	$-0.69 < \alpha \le -0.25$	$-0.39\alpha + 2.87$
	$-0.25 < \alpha \le -0.13$	2.77
	$-0.13 < \alpha \le 1$	$-0.50\alpha + 2.71$

Table 2.14 Dispersion of dense gas puffs

Concentration ratio (C_m/C_o)	Valid range for $\alpha = \log (g_o V_o^{1/3}/u^2)^{1/2}$	$\beta = \log [x/(V_o)^{1/3}]$
0.1	$\alpha \le -0.44$	0.7
	$-0.44 < \alpha \le 0.43$	$0.26\alpha + 0.81$
	$0.43 < \alpha \le 1$	0.93
0.05	$\alpha \le -0.56$	0.85
	$-0.56 < \alpha \le 0.31$	$0.26\alpha + 1.0$
	$0.31 < \alpha \le 1$	$-0.12\alpha + 1.12$
0.02	$\alpha \le -0.66$	0.95
	$-0.66 < \alpha \le 0.32$	$0.36\alpha + 1.19$
	$0.32 < \alpha \le 1$	$-0.26\alpha + 1.38$
0.01	$\alpha \le -0.71$	1.15
	$-0.71 < \alpha \le 0.37$	$0.34\alpha + 1.39$
	$0.37 < \alpha \le 1$	$-0.38\alpha + 1.66$
0.005	$\alpha \le -0.52$	1.48
	$-0.52 < \alpha \le 0.24$	$0.26\alpha + 1.62$
	$0.24 < \alpha \le 1$	$0.30\alpha + 1.75$
0.002	$\alpha \le 0.27$	1.83
	$0.27 < \alpha \le 1$	$-0.32\alpha + 1.92$
0.001	$\alpha \le -0.10$	2.075
	$-0.10 < \alpha \le 1$	$-0.27\alpha + 2.05$

Value-Time-Weighted Average (TLV-TWA) are very conservative estimates of work exposure. There are six alternate methods of toxic effect evaluation:

Method 1: Based on Emergency Response Planning (ERPG). This is formulated by American Industrial Hygiene Association. In this three ERPG values are used namely ERPG-1, ERPg-2, and ERPG-3.

Method 2: Based on the guidelines recommended by National Institute for Occupational Safety and Health (NIOSH), toxicity is evaluated. NIOSH recommends standards for Immediately Dangerous to Life and Health (IDLH) that explains the level of acceptable toxicity.

Method 3: Is based on the guidelines as recommended by National Research Council, Canada (NRC). NRC recommends Emergency Exposure Guidance Levels (EEGL) for different duration of exposure namely 1 hour EEGL and 24 hours EEGL.

Method 4: Is based on OSHA's Permissible Exposure Limits (PELs). This includes Occupational Safety and Health Administration, U.S. Department of Labor.

Method 5: Is based on the Environmental Protection Agency's (EPA) toxic end point. Guidelines recommended by EPA, U.S. EPA/6000/R-7/080 (2007): Sediment Toxicity Identification Evaluation guidelines are followed to estimate the toxicity.

Method 6: Is based on the guidelines recommended by American Conference of Governmental Industrial Hygienists (ACGIH). ACGIH [1994]. 1994–1995 recommends threshold limit values for chemical substances, physical agents, and biological exposure indices. Cincinnati, OH: American Conference of Governmental Industrial Hygienists.

2.21 Hazard Assessment and Accident Scenarios

The hazard assessment procedure commonly used in oil and gas industries is shown in Figure 2.15.

Development of accident scenario is given in Figure 2.16.

2.21.1 Damage Estimate Modeling: Probit Model

The probit value, in terms of probability units, is given by:

$$P = \frac{1}{(2\pi)^{1/2}} \int_{-\infty}^{Y-5} \exp\left(\frac{-u^2}{2}\right) du \qquad (2.19)$$

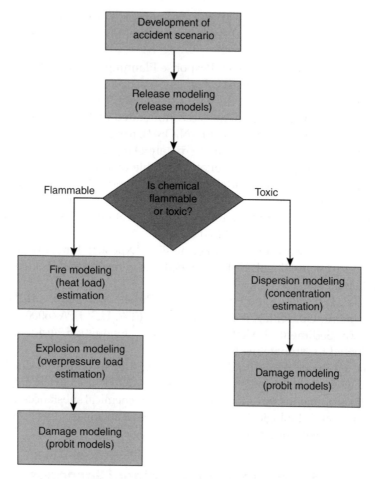

Figure 2.15   Hazard assessment

The probit function transforms the nonlinear dose response into a linear relationship as shown in Figure 2.17 (also given in Eq. (2.20)).

$$Y = K_1 + K_2 \ln (V) \tag{2.20}$$

where, K_1, K_2 are constants and V is the dose variable (due to over pressure, radiation, impulse, or concentration of dispersion). In a simplified form,

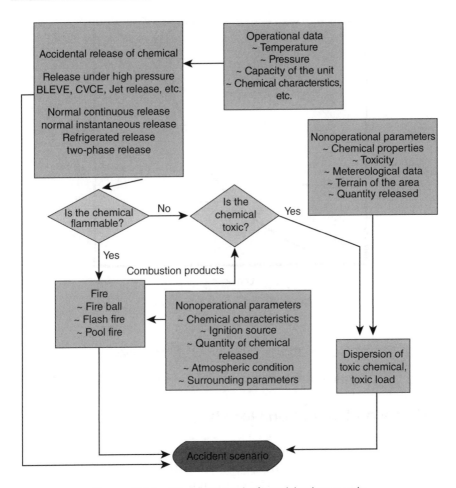

Figure 2.16 Development of accident scenario

probit value (Y) can be transformed to the percentage effect using the following relationship:

$$P = 50\left[1 + \frac{Y-5}{|Y-5|}\,\mathrm{erf}\left(\frac{|Y-5|}{\sqrt{2}}\right)\right] \tag{2.21}$$

where *erf* is error function.

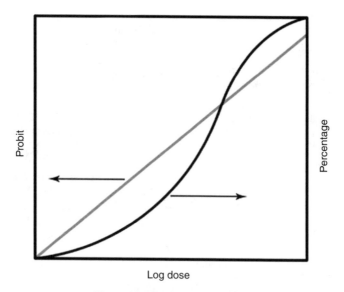

Log dose

Figure 2.17   Probit function

2.21.2 Probit Correlations for Various Damages

The probit correlations for various damages are given in Table 2.15.

2.22 Fire and Explosion Models

Dow Fire and Explosion Index (FEI) method is commonly used to study fire and explosion releases in oil and gas industries (Papazoglou et al., 2003). The flowchart to compute the FEI is given in Figure 2.18.

Step 1: Compute Material factor (MF)
Material factor depends on the vapor pressure and flammable (or explosive) characteristics.

Step 2: Compute factor F1
It depends on general process hazards. For example, hazards arising due to unit operation such as reaction, material handling, etc.

Step 3: Compute factor F2
It depends on the special process hazards. For example, hazards surrounding the unit that arise due to special conditions in operation.

Table 2.15 Probit correlations for various damages

Type of damage	Dose variable	Probit equation constants	
		K_1	K_2
Fire			
Burn deaths from fire	$(t^*I^{4/3})/10^4$	−14.9	2.56
Explosions			
Deaths from lung hemorrhage	P^0	−77.9	6.91
Eardrum rupture	P^0	−15.6	1.93
Structural damage	P^0	−23.8	2.92
Glass breakage	P^0	−18.1	2.79
Death from overpressure impulse	J	−46.1	4.82
Injuries from overpressure impulse	J	−39.1	4.45
Injures from flying fragments	J	−27.1	4.26
Toxic release and dispersion			
Death due to ammonia dose	$C^{2.0}*T$	−35.9	1.85
Death due to sulfur dioxide dose	$C^{1.0}*T$	−15.67	1
Death due to chlorine dose	$C^{2.0}*T$	−8.29	0.92
Death due to ethylene oxide dose	$C^{1.0}*T$	−6.19	1
Death due to phosgene dose	$C^{1.0}*T$	19.27	3.69
Death due to toluene dose	$C^{2.5}*T$	−6.79	0.41

T is time (s); I is radiation intensity (W/m²); P^0 is peak over pressure (N/m²); J is impulse (Ns/m²); C is exposed concentration (ppm); and T is duration of exposure (min).

Step 4: Compute Process Unit Hazard (PUH) using the following relationship:

$$PUH = (1 + F1) * (1 + F2) \tag{2.22}$$

Step 5: Compute FEI
The dow FEI is given by:

$$FEI = MF * PUH \tag{2.23}$$

Based on the FEI value computed, one can rate the degree of hazard as given in Table 2.16.

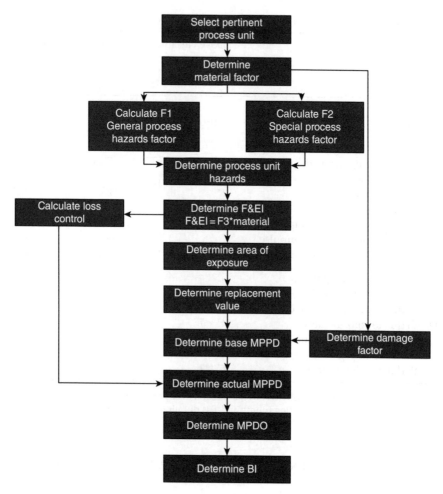

Figure 2.18 Flowchart to compute FEI

Table 2.16 The degree of hazard for different dow FEI

Dow FEI	Degree of hazards
1–60	Light
61–96	Moderate
97–127	Intermediate
128–158	Heavy
159 and above	Severe

Exercises 2

1. What are the main hazards related to oil and gas industry?

   Safety and injury hazards, health and illnesses hazards.

2. Write a short note on occupational safety and health management system.

   The insinuation of implementing an occupational safety and health management system at all workplaces came into limelight, when "Global Strategy on Occupational Safety and Health: Conclusions" were adopted by the "International Labour Conference" at its 91st session, 2003. The strategy advocates the application of a systems approach to the management of national OSH systems. Also, guidelines on Occupational Safety and Health Management Systems (ILO-OSH 2001) provide national/organizational framework for OSH management systems. As per these guidelines, the OSH management system should contain the main elements of policy, organizing, planning and implementation, evaluation and action for improvement.

3. What are the key features that should be fulfilled by an efficient safety and health management system?

   It should ensure safety of different operational sites by correctly mapping the business processes, risks, and controls involved in all the three segments (upstream, midstream, and downstream) of oil and gas industry. It should enable workers to follow consistent health and safety practices. It should help in managing site inspections, permits, violations, lessons learned, and best practices execution for oil and gas sector. It must be well documented (strategies and action plans) and should be easily understood and readily available to all the workers.

4. Name some of the components of an effective occupational safety and health management system.

   Health and safety plan, administration, work area management, H&S risk management, inventory management, etc.

5. What are the benefits of occupational safety and health management system?

   It enables oil and gas industry in performing hazard identification, risk assessment, and implementing various control methods, It ensures well-being of all the employees and thus contributes to a more inspired, and performance-driven workforce. Regular risk assessment process helps in frequent tracking and monitoring of health and safety indicators (both leading and lagging). Reduced costs associated with accidents and incidents improved regulatory compliance implementation of OSH management system gives competitive edge and improves relationships between stakeholders, such as clients, contractors, subcontractors, consultants, suppliers, employees, and unions.

6. What are the different safety measures in design and process operation employed in oil and gas industries?

   Inerting, explosion, fire prevention, sprinkler systems.

7. What are the different conditions that must be satisfied to cause fire accidents?

   Presence of combustive or explosive material, presence of oxygen to support combustion reaction, source of ignition to initiate the reaction.

8. What are the different fire and explosion control measures?

   Use explosion-proof equipments and instruments, use well-designed sprinkler systems, use modern design features.

9. What is the significance of inerting and purging methods?

   Reduce oxygen or fuel concentration below the target value, usually it is 4% below the limiting oxygen concentration, nitrogen, carbon dioxide, and others can be used, nitrogen is commonly used.

10. What are the different purging methods available in process industries?

    Vacuum purging, pressure purging, combined purging, vacuum and pressure purging with impure nitrogen, sweep-through purging, siphon purging.

11. How flammability diagrams are helpful in reducing the fire hazards?

    It determines whether or not flammable mixture exists, it also provides target concentration for inerting and purging, two distinct uses are placing vessel out of service and placing vessel into service.

12. What is meant by placing vessel out of service and placing vessel into service?

Placing vessel out of service: Gas concentration at points R and M are known from the flammability diagram of the fuel (e.g., methane), find the composition at point S, graphically placing vessel into service: gas concentration at points R and M are known, composition at point S can be determined graphically, nitrogen is pumped in till the point S is reached.

13. What are the different types of sprinkler systems?

Antifreeze sprinkler system, deluge sprinkler system, dry pipe sprinkler system, wet pipe sprinkler system.

14. Discuss about the ventilation guidelines used inside storage area.

System should be interlocked with sound alarm when ventilation fails, inlet and exhausts should be located to provide air movement across entire area, recirculation of air is permitted, stopped when air concentration >25% of lower flammability limit.

15. What are the major problems and risks involved in drilling operations?

Drill pipe sticking and pipe failure, lost circulation, hole deviation and borehole instability, mud contamination, formation damage, drill bit failure.

16. Name some of the quantitative risk analysis software used in the process industries.

Safeti, Phast risk, Risk, Risk spectrum, ASAP, Plato

17. Name some of the standards used for atmospheric storage tanks. Discuss any one in detail.

API 650, API 620, ASME sec V, ASTM, etc.

18. What are the factors which makes the management systems rigorous?

Complexity, hazard and risk, resource demands/availability, culture

19. Write a short note on process hazard analysis.

The process hazard analysis is a thorough, orderly, systematic approach for identifying, evaluating, and controlling the hazards of processes involving highly hazardous chemicals. The employer must perform an initial process hazard analysis (hazard evaluation) on all processes covered by this standard. The process hazard analysis methodology selected must be appropriate to the complexity of the process and must identify, evaluate, and control the hazards involved in the process.

20. Differentiate between OSHA's process safety management regulations and EPA risk management program.

PSM: Protects the workforce, protects contractors, protects visitors to the facility, basically protects the workplace

RMP: Protects the community, protects the general public around the facility, protects adjacent facilities such as schools and hospitals.

3

Accident Modeling, Risk Assessment, and Management

3.1 Introduction

Toxicology is defined as a qualitative and quantitative study of the adverse effects of toxicants on biological organisms. A toxicant can be a chemical or physical agent, including dusts, fibers, noise, and radiation. Toxicants enter the human body through any of the following: (i) ingestion, which is through the mouth into stomach; (ii) inhalation, which is through the mouth or nose into lungs; (iii) injection, which is through the cuts into skin; and (iv) dermal absorption, which is through the skin membrane.

3.2 Dose Versus Response

Biological organisms respond differently to the same dose of toxicants. Factors responsible for such variation in their response are age, sex, weight, diet, and general health conditions. Figure 3.1 shows the response behavior for the dose of the toxicants. Figure 3.2 shows the logarithmic plot of the dose–response behavior under the influence of toxicants.

Health, Safety, and Environmental Management in Offshore and Petroleum Engineering, First Edition.
Srinivasan Chandrasekaran.
© 2016 John Wiley & Sons, Ltd. Published 2016 by John Wiley & Sons, Ltd.
Companion website: www.wiley.com/go/chandrasekaran/hse

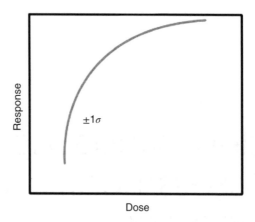

Figure 3.1   Dose vs. response behavior of toxicants

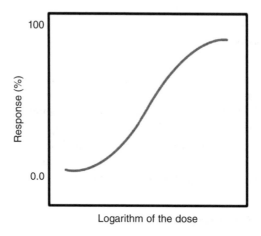

Figure 3.2   Logarithmic dose vs. response behavior of toxicants

3.2.1 Various Types of Doses

Different types of doses are briefly discussed next and the plot is shown in Figure 3.3.

(a) Lethal Dose (LD)

If the response to the chemical or agent is lethal and deadly, the response versus log dose curve is called LD curve.

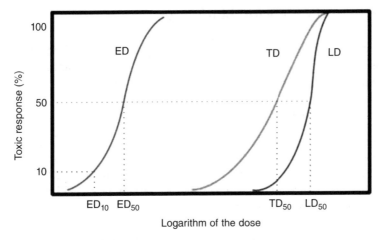

Figure 3.3   Various types of doses

(b) Effective Dose (ED)

If the response to the chemical or agent is minor and reversible (e.g., minor eye irritation, skin infection, sour throat, eye burning sensation etc.), the response versus log dose curve is called ED curve.

(c) Toxic Dose (TD)

If the response to the agent is toxic (i.e., it causes an undesirable response that is not lethal but irreversible, e.g., liver or lung damage), the response–log dose curve is called TD curve.

(d) Lethal Concentration (LC)

For gas concentration, logarithm of the dose is used.

3.2.2 Threshold Limit Value (TLV) Concentration

The TLV concentration is based on 760 Hg pressure at 25°C and a molar volume of 24.45 l. The equations for converting parts per million (ppm) to mg/m³ and vice versa are as follows:

$$\text{TLV in mg/m}^3 = \frac{(\text{TLV in ppm}) \times (\text{gram molecular weight of substance})}{24.45}. \tag{3.1}$$

$$\text{TLV in ppm} = \frac{(\text{TLV in mg/m}^3) \times 24.45}{(\text{gram molecular weight of substance})}. \tag{3.2}$$

3.3 Fire and Explosion Modeling

Chemical process systems contain substantial hazard that arise from fire and explosions (Ramamurthy, 2011). Three common chemical plant accidents are fire, explosion, and toxic releases (Chamberlain, 1987).

3.3.1 Fundamentals of Fire and Explosion

(a) Fire

Fire is a rapid exothermal oxidation of ignited fuel. Fuel can be in solid, liquid, or vapor form. Fire will release energy in the form of thermal radiation, which takes time to reach its peak intensity. Fire can also be formed as a result of explosion.

(b) Explosion

Explosion is a rapid expansion of gases resulting from (rapidly moving) pressure or shock waves. Expansion can be mechanical or resulting from a chemical reaction. Explosion damage is caused by the pressure or shock waves, which release energy rapidly that can cause fire (Chuan-Jie et al., 2013).

(c) Accident Prevention

Fire and explosion accidents can be prevented using the knowledge of fire and explosion characteristics of materials, nature of fire, and the explosion process. The procedure to reduce fire and explosion hazards is by using fire triangle as shown in Figure 3.4. Fire and explosion can be prevented by removing any one of the arms of the fire triangle.

3.4 Fire and Explosion Characteristics of Materials

Fire characteristics of flammable materials are as follows:

(a) Auto-ignition temperature (AIT)

It is the fixed temperature above which the material may not require any external ignition source for combustion.

(b) Flash point

It is the lowest temperature at which liquid gives up enough vapor to maintain continuous flame.

(c) Flammability limits

It is the range of vapor concentration that could cause combustion on meeting the ignition source. There are two limits in which the fuel will

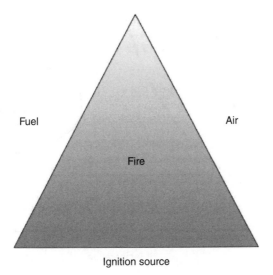

Figure 3.4   Fire triangle

catch fire: Lower Flammability Limit (LFL) and Upper Flammability Limit (UFL). LFL is the limit below which the mixture will not burn due to lean mixture and UFL is the limit above which the mixture will not catch fire as the mixture is too rich to catch fire.

(d) Limiting oxygen concentration (LOC)

It is the minimum oxygen concentration below which combustion is not possible, with any fuel mixture. It is expressed as volume percentage of oxygen. It is also called as Minimum Oxygen Concentration (MOC) or Maximum Safe Oxygen Concentration (MSOC).

(e) Shock wave

These are abrupt pressure waves moving through a medium. A shock wave in open air is usually followed by wind, which is called a blast wave. One of the important characteristics of the shock wave is that the pressure increase in shock wave is so rapid that the process is mostly adiabatic.

(f) Over pressure

It is pressure that is imparted on an object by a shock wave.

Figure 3.5 shows the flammability characteristics of materials.

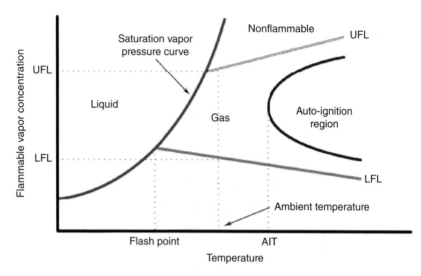

Figure 3.5 Flammability characteristics of materials

Table 3.1 Flammability constants

Chemical group	a	b	c
Hydrocarbons	225.1	537.6	2217
Alcohols	225.8	390.5	1780
Amines	222.4	416.6	1900
Acids	323.2	600.1	2970
Ethers	275.9	700.0	2879
Sulfur	238.0	577.9	2297
Esters	260.5	449.2	2217
Halogens	262.1	414.0	2154
Aldehydes	264.5	293.0	1970

3.4.1 Flammability Characteristics of Liquids

Flammability characteristics of liquids are experimentally determined by open-cup method. Flammability characteristics of the liquid can be expressed using the following relationships where the flammability constants (a, b, c) are given in Table 3.1. T_f is flash point, T_b is boiling point in Kelvin.

$$T_f = a + \frac{b\left(c / T_b\right)^2 e^{-(c/T_b)}}{\left(1 - e^{-(c/T_b)}\right)^2} \tag{3.3}$$

where, T_f is flash point, T_b is boiling point in Kelvin.

3.4.2 Flammability Characteristics of Vapor and Gases

Flammability limits for vapor are determined experimentally in a specially closed vessel apparatus. Flammability limits for mixture of gases and vapors are given by the following relationship:

$$LFL_{mixture} = \frac{1}{\sum_n y_i / LFL_i} \tag{3.4}$$

$$UFL_{mixture} = \frac{1}{\sum_n y_i / UFL_i} \tag{3.5}$$

where, LFL_i is the LFL for the ith component (in volume %) fuel and air; y_i is the mole fraction of ith component on a combustible basis; n is the number of combustible species.

3.5 Flammability Limit Behavior

The flammability limit behavior depends on the pressure and temperature. When the temperature increases, UFL increases, while the LFL decreases; this results in an increase in the flammability range. As the pressure increases, UFL increases, while increase in pressure has no significant effect on the LFL. The relationship between UFL with that of pressure is:

$$UFL_p = UFL + 20.6\left(\log P + 1\right) \tag{3.6}$$

where, P is absolute pressure in MPa.

3.6 Estimation of Flammability Limits Using Stoichiometric Balance

For many hydrocarbon vapors, LFL and UFL are functions of stoichiometric concentration (C_{st}) of the fuel. For a general combustion reaction, following relationship holds good:

$$C_m H_x O_y + z O_2 \rightarrow m CO_2 + \left(\frac{x}{2}\right) H_2 O \tag{3.7}$$

Stochiometric concentration is given by:

$$C_{st} = \frac{21\%}{0.21 + z} \tag{3.8}$$

where, z is calculated using the following relationship:

$$z = m + \frac{1}{4}x + \frac{1}{2}y \tag{3.9}$$

Flammability limits can be now estimated using the following relationship:

$$\text{LFL} = 0.55 C_{st} \tag{3.10}$$

$$\text{UFL} = 3.5 C_{st} \tag{3.11}$$

3.6.1 Stoichiometric Balance

Stoichiometric balance gives the relationship between the quantities of substances that take part in a chemical (or combustion) reaction. This is typically a ratio of whole integers, which includes the calculation of quantitative relationships of the reactants and products in a balanced chemical reaction. LFL and UFL concentrations are determined using the following relationships:

$$\text{LFL} = \frac{55}{4.76m + 1.19x - 2.38y + 1} \tag{3.12}$$

$$\text{UFL} = \frac{350}{4.76m + 1.19x - 2.38y + 1} \tag{3.13}$$

Alternatively, flammability limits can also be determined using the following relationships:

$$\text{LFL} = \frac{-3.42}{\Delta H_c} + 0.569 \Delta H_c + 0.0538 \Delta H_c^2 + 1.80 \tag{3.14}$$

$$\text{UFL} = 6.30 \Delta H_c + 0.567 \Delta H_c^2 + 23.5 \tag{3.15}$$

3.6.2 Estimation of Limiting Oxygen Concentration (LOC)

LOC has units of percentage of moles of oxygen in total moles. For hydrocarbons, LOC is estimated using stoichiometry relationship of the combustion reaction and the LFL is:

$$\text{LOC} \sim (z) \cdot (\text{LFL}) \tag{3.16}$$

3.7 Flammability Diagram for Hydrocarbons

Flammability diagram determines whether or not a given mixture is flammable. This will be helpful to control or prevent fire and explosion of flammable mixtures. Flammability diagram depends on the chemical species and is a function of temperature and pressure. A typical flammability diagram is shown in Figure 3.6.

3.7.1 Constructing Flammability Diagram

The construction of flammability diagram is done using the following steps:

- Mark three arms of the triangle and mark them as oxygen, nitrogen, and fuel arms respectively. Apex and origin of these arms are marked in an anti-clockwise order as shown in Figure 3.6.
- Draw the air line by connecting apex of the fuel arm with that of 79% of nitrogen. This is called as an air line.
- Draw LFL and UFL of the fuel mixture on the air line (% fuel in air).
- Locate the stoichiometric point on the oxygen axis.
- Draw a stoichiometric line from this point to 100% nitrogen apex.
- Locate LOC on the oxygen axis and draw a line parallel to the fuel arm until it intersects the stoichiometric line.
- Draw a point at this intersection.
- Draw LFL and UFL in pure oxygen, if known (% of fuel in pure oxygen).
- Connect the points to get the flammability diagram, as shown in the figure.

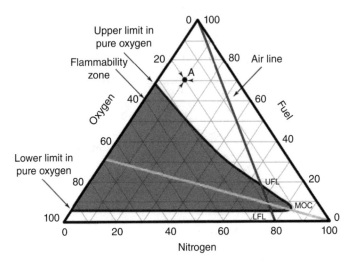

Figure 3.6 Flammability diagram

Example

The flammability characteristics of methane (CH_4) are as follows:

Flammability limit in air: LFL = 5.3% fuel in air
UFL = 15% fuel in air.
Flammability limit in pure oxygen: LFL = 5.1% fuel in oxygen
UFL = 61% fuel in oxygen.
Limiting oxygen concentration (LOC) = 12% oxygen.

Let us construct the flammability diagram for methane.
The general equation of combustion is given by:

$$C_m H_x O_y + zO_2 \longrightarrow mCO_2 + (x/2) H_2O$$

Balanced combustion reaction is given by:

$$CH_4 + 2O_2 \longrightarrow CO_2 + 2H_2O$$

By comparing the above two equations, $z = 2$. Stoichiometric point is given by:

$$[z/(1+z)] \times 100 = 66.7\% \text{ of oxygen}$$

The corresponding flammability diagram is shown in Figure 3.7.

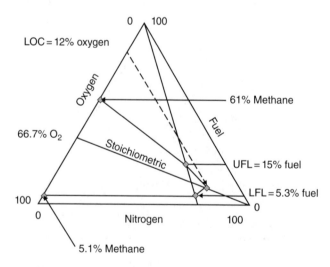

Figure 3.7 Flammability diagram for methane

3.8 Ignition Energy

The Minimum Ignition Energy (MIE) required to initiate combustion reaction of hydrocarbons also controls the probability of fire in process industries. All flammable materials have MIE, which depends on its (composition) or mixture, concentration, temperature, and pressure. MIE for some of the chemicals is given in Table 3.2. Ignition can be initiated through many ways. Some of the common initiating ignition sources are given Table 3.3.

Table 3.2 Minimum ignition energy for some chemicals

Chemical	MIE (mJ)
Acetylene	0.02
Benzene	0.225
Butadiene	0.125
Butane	0.260
Hexane	0.248
Ethane	0.24
Ethene	0.124
Hydrogen	0.018
Methane	0.28
Propane	0.25

Table 3.3 Ignition sources

Source	%
Electric	23
Smoking	18
Friction	10
Overheated material	8
Hot surfaces	7
Flames	7
Sparks	5
Others	22

3.9 Explosions

Explosion is a rapid release of energy causing development of pressure or shock waves. There are different types of industrial explosions as discussed next:

Confined Vapor Cloud Explosion (CVCE):
It is a type of explosion that happens in a vessel or in a confined space (e.g., inside a building). It is generally caused due to the release of high pressure or chemical energy.

Vapor Cloud Explosion (VCE):
It is a type of explosion caused by the instantaneous vapor cloud, which is formed in air due to the release of flammable chemicals into atmosphere.

Boiling Liquid Expanding Vapor Explosion (BLEVE):
BLEVE is caused due to the instantaneous release of a large amount of vapor through narrow opening under the pressurized condition (Tasneem Abbasi and Abbasi, 2007).

Vented Explosion (VE)
It is caused due to venting of chemicals at a high velocity.

Dust Explosion
Dust explosion is caused due to the rapid combustion of fine solid particles.

3.10 Explosion Characteristics

Explosion energy is dissipated in different forms: (i) pressure wave; (ii) projectiles; (iii) thermal radiation; and (iv) acoustic energy (Planas-Cuchi et al., 2004). Some of the types of energy, which is generated due to explosion are discussed next:
   Blast wave is a shock wave in open air, which is generally followed by a strong wind. *Overpressure* of an object is a result of an impacting shock waves on any object. *Detonation* is a kind of explosion in which reaction front moves at a speed greater than that of sound in the given medium. *Deflagration* is a kind of explosion in which reaction front (energy front) moves at a lesser speed than that of sound in the given medium.

3.11 Explosion Modeling

Explosions result in a blast or a pressure wave moving out from the explosion center at the speed of sound. Shock wave or overpressure is the basic cause for damages during explosions. Missiles or projectiles are other important

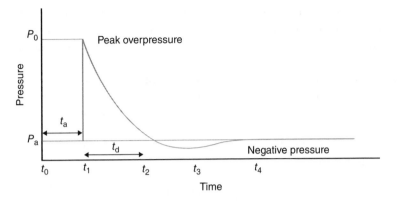

Figure 3.8 Variation of overpressure with time

sources of damage. Generally, damage caused by explosions is a function of rate of pressure increase and the duration of blast wave. As explosion generates a rapid rise in pressure, damage caused by blasts is estimated based on the peak side-on overpressure. The propagating wave causes damage to the objects located on its path, which is then followed by a negative pressure wave. This causes further damage before the pressure waves return to the atmospheric pressure. Therefore, damage depends upon various factors: (i) maximum pressure reached; (ii) velocity of propagation; and (iii) environmental characteristics. The variation of overpressure with time due to blast wave is given in Figure 3.8. The area under the curve is the measure of severity of explosion.

3.12 Damage Consequences of Explosion Damage

One of the most common methods used to determine the consequences that arise from the explosion damage is Tri Nitro Toluene (TNT) equivalence method (Pasman et al., 2009). TNT is an important explosive, which can quickly change its form from solid to the hot expanding gas (Aven and Vinnem, 2007; Pasman et al., 2009). After explosion, it produces *soot*, which is a black powder. The chemical reaction for it is:

$$2C_7H_5N_3O_6(s) \rightarrow 3N_2 + 7CO(g) + 5H_2O(g) + 7C(s) \qquad (3.17)$$

As TNT contains elements of carbon, oxygen, and nitrogen, it produces highly stable substances with a strong bonding between them when it burns.

TNT explosions are chemically unstable, which implies that it does not require much force to break their bond. Steps involved in computing equivalent mass of TNT and the scaled distance are discussed in the following:

Step 1: Determine the total mass of fuel involved (m).
Step 2: Determine the energy of explosion (ΔH_c).
Step 3: Estimate the energy of explosion, η, which usually varies from 1 to 15%.
Step 4: Calculate the equivalent mass of TNT using the following relationship:

$$m_{TNT} = \frac{\eta \, m \, \Delta H_c}{E_{TNT}} \qquad (3.18)$$

where, E_{TNT} is the energy of explosion of TNT (=4686 kJ/kg)
Step 5: Determine the scaled distance (Z_e) using the following relationship:

$$Z_e = \frac{r}{\sqrt[3]{m_{TNT}}} \qquad (3.19)$$

where, r is the distance from the explosion site to the point of concern (in m).

The resulting overpressure P_o can be calculated either graphically or mathematically using the following relationship. Graphically, overpressure P_o can be calculated using Equation (3.20) in which the scaled overpressure P_s can be determined using Figure 3.9, where P_a is the atmospheric pressure (Table 3.4).

$$P_o = P_s \times P_a \qquad (3.20)$$

$$P_o = P_a \left[\frac{1616\left[1+\left[Z_e / 4.5\right]^2\right]}{\sqrt{1+\left[Z_e / 0.048\right]^2} \times \sqrt{1+\left[Z_e / 0.32\right]^2} \times \sqrt{1+\left[Z_e / 1.35\right]^2}} \right] \qquad (3.21)$$

Alternative methods are also seen in the literature based on the degree of congestion or confinements. See, for example, TNO Multienergy model and Baker–Strehlow model.

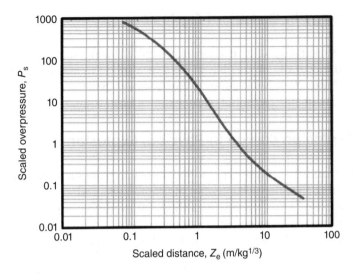

Figure 3.9 Scaled overpressure

Table 3.4 Explosion damages caused by overpressure

Overpressure (kPa)	Damage
0.28	Loud noise (143 dB)—glass failure
0.69	Breakage of small windows
2.07	Safe distance (probability of 0.95 of no serious damage below this value)
3.4–3.6	Windows shatter; occasional damage to window frames
4.8	Minor damage to house structure
6.9–13.8	Significant damage to wooden and asbestos
15.8	Lower limit of serious damage
17.2	Destruction of brick houses
27.6	Cladding of industrial building ruptures, oil tank ruptures, 50% probability of human fatality
34.5–48.2	Nearly complete destruction of houses
62	Loaded trains completely gets damaged
68.9	Probable total destruction of buildings, heavy machinery
75	90% probability of human fatality; concrete and steel structures completely gets damaged, etc.

3.13 Energy in Chemical Explosions

Energy released during a reaction is computed using standard thermodynamics. The heat of combustion is used as a mode to assess the explosion strength. From the past studies, it can be seen that the explosion energy differs by about 10% from that of the value of heat of combustion.

3.14 Explosion Energy in Physical Explosions

In the mechanical or physical explosions, a reaction does not occur. Energy is obtained from the energy content of the contained substance. Four common approaches used are: (i) Brode's method; (ii) isentropic expansion method; (iii) isothermal expansion method; and (iv) laws of thermodynamics. Equations (3.22–3.24) are used in place of the (first) three methods:

$$E_{\text{Brode}} = \frac{(P_2 - P_1)\,V}{\gamma - 1} \tag{3.22}$$

$$E_{\text{isentopic}} = \frac{P_2\,V}{\gamma - 1}\left[1 - \left[\frac{P_1}{P_2}\right]^{\frac{\gamma - 1}{\gamma}}\right] \tag{3.23}$$

$$E_{\text{isothermal}} = P_2 V \ln\left(\frac{P_2}{P_1}\right) \tag{3.24}$$

where, E is the explosion energy; P_1 is the atmospheric pressure, P_2 is the bursting pressure, V is the volume of vessel, and γ is the heat capacity ratio.

3.15 Dust and Gaseous Explosion

Gas molecules are smaller and well-defined in size, whereas dust particles vary in size; this variation in magnitude is larger than molecules. K_G and K_{st} are deflagration index for gas and dust, respectively, which are given by Equations (3.25) and (3.26), respectively. Variation in pressure with respect to time for gas and dust explosion is given in Figure 3.10.

$$\left(\frac{dp}{dt}\right)_{\text{max}} \sqrt[3]{V} = K_{st} \tag{3.25}$$

$$\left(\frac{dp}{dt}\right)_{\text{max}} \sqrt[3]{V} = K_G \tag{3.26}$$

3.16 Explosion Damage Estimate

Explosion generates a rapid rise in pressure, which produces shock waves. As wave propagates, it encounters damage in its path, which is then followed by a negative pressure wave. This causes further damage before it returns to atmospheric pressure. Thus, damage caused by dust and gas explosions depends on the maximum pressure reached, velocity of propagation, and other environmental characteristics; variation of pressure with time is given in Figure 3.11. While deflagration can cause rise in pressure up to eight times that of the initial, detonation process can cause a rise even more. Rate of pressure rise depends on the characteristics of the mixture and the degree of containment of explosives (Bubbico and Marchini, 2008).

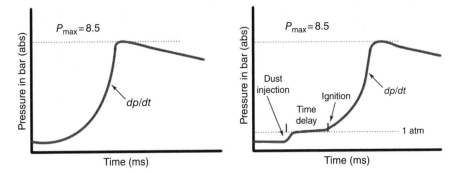

Figure 3.10   Variation of pressure for gas and dust with regard to time

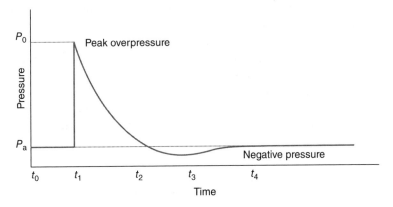

Figure 3.11   Pressure variation with respect to time

To prevent fire and explosion of any flammable mixture, it is important to reduce the ignition sources or ensure other safer design procedures. They include (i) reduction of inventories; (ii) substitution with less dangerous materials; and (iii) reduction of operational temperature and pressure.

3.17 Fire and Explosion Preventive Measures

Some of the strategies that are commonly used in oil and gas industries are discussed in the next subsection:

3.17.1 Inerting and Purging

This procedure aims to reduce oxygen or fuel concentration lower than a target value using an inert gas. While nitrogen, carbon dioxide, and others are possible choice of inert gases, nitrogen is commonly used at a control point of 4% below LOC. Some of the purging methods are vacuum purging, pressure purging, combined purging, vacuum and pressure purging with impure nitrogen, sweep-through purging, and siphon purging. During purging methods, it is assumed that pure nitrogen is purged, which mixes well inside the vessel and therefore ideal gas behavior.

3.17.1.1 Vacuum Purging

It is one of the most commonly used inerting procedures for vessels. In vacuum purging, the vessel is evacuated and replaced with inert gas. The procedure includes drawing vacuum from the vessel and then replace it with inert gas. This cycle is repeated until the desired concentration is reached. The cycle is shown in Figure 3.12.

Initial oxidant concentration under vacuum (y_0) is the same as the initial concentration. Number of moles at initial high pressure (P_H) and initial low pressure (P_L) are computed using an initial equation of state, for an ideal gas behavior. n_H and n_L are total moles in the atmosphere and vacuum states, respectively. At point A in the figure, the number of oxidants is calculated using Equation (3.27) and that at point B, using Equation (3.28).

$$\text{at A}, n_{\text{oxygen}} = y_0 \left[\frac{P_L V}{R_g T} \right] \tag{3.27}$$

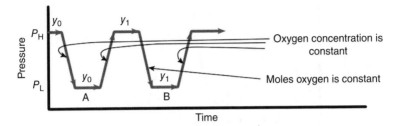

Figure 3.12   Variation of pressure due to vacuum purging

$$\text{at B}, n_{\text{total}} = \left[\frac{P_H V}{R_g T}\right] \tag{3.28}$$

The number of moles for oxidants is calculated using Dalton's law. At the end of the first cycle, new oxidant (lower) concentration is y_1, where y_1 is the oxygen concentration after the first purge with nitrogen. Similarly at the end of the second cycle, oxygen concentration is given by y_2, as given in Equation (3.30).

$$y_1 = \frac{n_{\text{oxygen}}}{n_{\text{total}}} = \frac{y_0 \left[P_L V / R_g T\right]}{P_H V / R_g T} = y_0 \frac{P_L}{P_H} \tag{3.29}$$

$$y_2 = y_1 \frac{P_L}{P_H} = y_0 \frac{P_L}{P_H}\left[\frac{P_L}{P_H}\right] = y_0 \left[\frac{P_L}{P_H}\right] \tag{3.30}$$

For the ith cycle, the oxygen concentration is given by Equation (3.31).

$$y_i = y_0 \left(P_L / P_H\right)^i \tag{3.31}$$

In this process, it is assumed that the total mass added to each cycle is constant. For ith cycles, total nitrogen gas moles is given by Δn_{N_2}, which is calculated using Equation (3.32).

$$\Delta n_{N_2} = i\left(P_H - P_L\right)\frac{V}{R_g T} \tag{3.32}$$

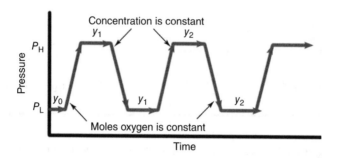

Figure 3.13   Pressure cycle due to pressure purging

3.17.1.2 Pressure Purging

Vessels can be pressure-purged by adding inert gas under pressure. After this added gas is diffused throughout the vessel, it is vented to atmospheric pressure for flushing. The cycle is shown in Figure 3.13. Initial oxygen concentration in the vessel (y_0) is computed after the vessel is pressurized. Number of moles for this pressurized state is n_H and that of atmospheric is n_L. Oxygen concentration for the ith cycle is given by Equation (3.31).

3.17.1.3 Combined Pressure–Vacuum Purging

In this method, both pressure and vacuum purging are carried out to purge a vessel. Computational procedure depends on whether the vessel is first evacuated or pressurized. If the vessel is pressurized first, then the beginning of the cycle is defined as the end of the initial pressurization; a variation is given in Figure 3.14. Oxygen mole fraction at this stage is the same as that of the initial mole fraction. Remaining cycles are identical to that of the pressure purge operation. If the initial oxygen mole fraction is 0.21, then the oxygen mole fraction at the end of this initial pressurization is given by:

$$y_0 = 0.21 \left[\frac{P_0}{P_H} \right] \tag{3.33}$$

Let i be # of cycles after the initial pressurization, then for $(i+1)$th cycle, oxygen mole fraction is given by:

$$y_i = y_0 \left[\frac{n_L}{n_H} \right]^i = y_0 \left[\frac{P_L}{P_H} \right] \tag{3.34}$$

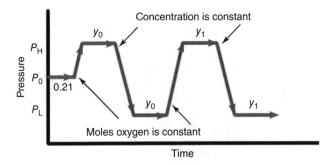

Figure 3.14 Pressure cycle due to combined pressure vacuum purging

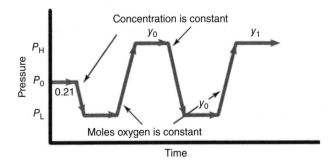

Figure 3.15 Pressure cycle if the evacuation is done first

If the evacuation is done first, then the beginning of the cycle is defined as the end of the initial evacuation as shown in Figure 3.15. Oxygen mole fraction at this state is the same as that of the initial mole fraction. Remaining cycles are identical to that of the vacuum purge operation. For $i+1$th cycle, oxygen concentration is given by:

$$y_i = y_0 \left[\frac{n_L}{n_H} \right]^i = y_0 \left[\frac{P_L}{P_H} \right] \tag{3.35}$$

3.17.1.4 Pressure and Vacuum Purging with Impure Purging

For a pressure purging, total moles of oxygen at the end of first pressurization is given as the sum of moles that are initially present and the moles included with the nitrogen, which is given by:

$$n_{oxy} = y_0 \left[\frac{P_L V}{R_g T} \right] + y_{oxy} \left(P_H - P_L \right) \left[\frac{V}{R_g T} \right] \tag{3.36}$$

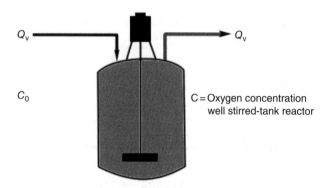

Figure 3.16   Sweep-through purging

Mole fraction of oxygen at the end of the first cycle is given by:

$$y_1 = \frac{n_{oxy}}{n_{total}} = y_0 \left[\frac{P_L}{P_H} \right] + y_{oxy} \left[1 - \frac{P_L}{P_H} \right] \tag{3.37}$$

Generalizing the above equation, oxygen concentration at the end of *i*th pressure cycle is given by:

$$y_1 = y_{i-1} \left[\frac{P_L}{P_H} \right] + y_{oxy} \left[1 - \frac{P_L}{P_H} \right]$$

$$\left(y_1 - y_{i-1} \right) = \left[\frac{P_L}{P_H} \right]^i \left(y_0 - y_{oxy} \right) \tag{3.38}$$

3.17.1.5 Comparison of Pressure and Vacuum Purging

Due to greater pressure difference, the pressure purging is faster. Vacuum purging uses lesser inert gas than that of pressure purging as the oxygen concentration is reduced primarily by vacuum.

3.17.1.6 Sweep-Through Purging

This purging process adds purge gas into a vessel at one opening and withdraws the mixed gas from the vessel at atmosphere from another opening as shown in Figure 3.16. This is generally used when the vessel or equipment is not rated for pressure or vacuum. Mass balance on oxygen is given by:

$$V \frac{dC}{dt} = C_0 Q_v - C Q_v \tag{3.39}$$

Hence, volumetric quantity of inert gas required to reduce oxygen concentration from C_1 to C_2 is $Q_v t$, which is given by:

$$Q_v t = V \ln\left[\frac{C_1 - C_0}{C_2 - C_0}\right] \tag{3.40}$$

3.17.1.7 Siphon Purging

The sweep-through purging requires large quantities of nitrogen. This could be expensive when purging is done on a large storage vessel. In such cases, siphon purging is preferred. Purging process starts by filling the vessel with liquid, which is preferably water. Purged gas is subsequently added to the vapor space as liquid is drained from the vessel. Volume of the purge gas is equal to that of the volume of the vessel. Therefore, the rate of purging is equivalent to the volumetric rate of liquid discharge from the vessel.

3.18 Use of Flammability Diagram

Objective of the flammability diagram is to identify the flammable region of a given (mixture) hydrocarbon. It determines whether a flammable mixture exists and provides target concentration for inerting and purging. This can also be used to prevent fire hazards. Knowing the fuel concentration of the given composition, one can decide to place the vessel out of service or add the vessel into the service mode, as explained in the following sections.

3.18.1 Placing a Vessel Out of Service

Typical flammability diagram is shown in Figure 3.17. Gas concentration at points R and M are known. Composition at point S is given by:

$$OSFC = \frac{LFL}{1 - z[LFL / 21]} = \frac{LOC}{z\left[1 - [LOC / 21]\right]} \tag{3.41}$$

where, OSFC is the Out of Service Fuel Concentration at point S, LFL is the volume of percent of fuel in air percent of oxygen, LOC in volume percent of oxygen, and z is the stoichiometric oxygen coefficient, which is obtained from the combustion reaction.

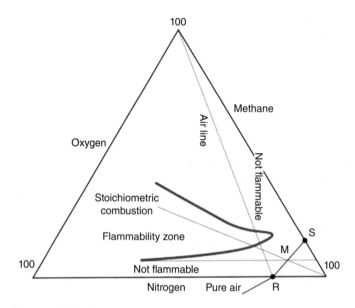

Figure 3.17 Flammability diagram for vessel out of service

3.18.2 Placing a Vessel into Service

Referring to Figure 3.18, composition at point S is given by Equation (3.42). Nitrogen concentration at point S is equal to 100-ISOC

$$\text{ISOC} = \frac{z \times \text{LFL}}{1 - [\text{LFL}/100]} = \frac{z \times \text{LOC}}{z - [\text{LOC}/100]} \tag{3.42}$$

where, ISOC is the In-Service Oxygen Concentration in volume percentage.

3.19 NFPA 69 Recommendations

National Fire Protection Association (NFPA) has set guidelines for fire safety practices. According to NFPA standards, target oxygen concentration for storage vessels should not exceed 2% below the measured LOC, if the oxygen content is continuously monitored. If LOC is lesser than 5%, then the target oxygen concentration should not exceed 60% of that of LOC. In addition, if the oxygen concentration is not continuously monitored, equipment must not operate more than 60% of LOC or 40% of LOC if LOC is below 5%.

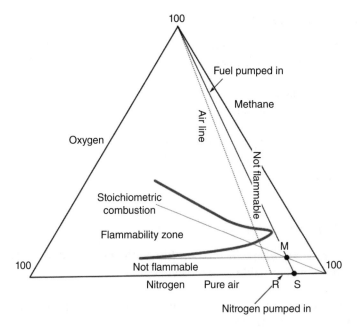

Figure 3.18   Flammability diagram for vessel in service

3.20 Explosion-Proof Equipments

Explosion-proof plants are designed not to prevent entry of flammable vapors or gases, but to withstand internal pressure and prevent combustion. For example, use of conduits with specially sealed connections around the junction boxes in wiring can be seen as common practice. Based on the area and material, they are classified into class systems, group systems, division systems.

3.20.1 Class Systems

Classes are related to the nature of flammable material.

Class I refers to the locations where flammable gases/vapors are present
Class II refers to the locations where flammable dust are present
Class III refers to the hazard locations where combustible fibers or dust are present but not likely to be in suspension.

3.20.2 Group Systems

Groups are based on the presence of specific types of chemicals. Chemicals having equivalent hazards are grouped as:

Group A: Acetylene
Group B: Hydrogen, ethylene
Group C: Carbon dioxide and hydrogen sulfide
Group D: Butane, ethane, ethyl alcohol
Group E: Aluminum dust
Group F: Carbon black
Group G: Flour

3.20.3 Division Systems

Divisions are categorized in relationship with the probability of material being within the flammable or explosive region.

Division 1 refers to the case where the probability of ignition is high and flammable concentration is normally present.
Division 2 refers to the case where hazards can occur only at abnormal conditions. Flammable materials are normally contained in a closed container or system.

3.21 Ventilation for Storage and Process Areas

3.21.1 Storage Areas

For storage areas, ventilation for inside space is recommended under certain conditions. For the rate of ventilation of $0.3\,m^3/min/m^2$ floor area, ventilation systems are to be interlocked to an alarm system as a mandate. When ventilation fails, location of the inlet and exhausts need to be made visible for free movement across the entire area. Recirculation of air is permitted for improving the dilution of the mixture; but stopped when air concentration exceeds 25% of LFL.

3.21.2 Process Areas

For process areas, a minimum of $0.3\,m^3/min/m^2$ floor area ventilation has to be provided. System has to be interlocked to an alarm when the ventilation fails. While ventilation standards are as same as that for the storage areas, ventilation system is designed to contain the concentrations within 1.5 m radius from all those sources whose LFL is lesser than 25%.

3.22 Sprinkler Systems

Sprinkler systems are also one of the effective means of controlling spread of fire. It is mandatory for all process plants to have a well-designed sprinkler system for ensuring safety to personnel and plants and equipments. A few sprinkler systems that are common in oil and gas industries are discussed in the next section:

3.22.1 Anti-freeze Sprinkler System

It consists of a wet-pipe system that contains an anti-freezing solution, which is connected to the water supply system. On demand, the sprinkler system opens the valves and pressurizes the liquid to control the fire hazard.

3.22.2 Deluge Sprinkler System

It consists of open sprinklers and an empty line that is connected to a water supply line through a valve. The valve is opened upon detection of heat and flammable material and water is sprayed through the nozzle of the sprinklers, which controls the spread of fire.

3.22.3 Dry Pipe Sprinkler System

This type of system is filled with nitrogen or air under pressure. When sprinkler nozzles are opened by heat, system is depressurized and allows water to flow into the system, which is further sprayed through the nozzles.

3.22.4 Wet Pipe Sprinkler System

This type of system contains water, which is discharged through nozzles that are opened when fire outbreaks in a plant. The design of sprinkler systems are based on the NFPA 69: Standards on Explosion prevention systems. As per NFPA norms, nominal discharge rate for 12.5 mm orifice spray nozzle are as shown in Table 3.5.

Table 3.5 Nominal discharge rate as per NFPA norms

Quantity of water (m³/s)	0.04	0.055	0.075	0.112	0.13
Pressure (N/mm²)	0.069	0.138	0.24	0.52	0.69

3.23 Toxic Release and Dispersion Modeling

3.23.1 Threshold Limit Values (TLVs)

TLVs represent conditions to which all workers will be repeatedly exposed every day without adverse health effects. For any dose value below this dosage, human body can be detoxified. The two agencies who established TLVs are The American Conference of Governmental Industrial Hygienists (ACGIH) and Occupational Safety and Health Administration, U.S. (OSHA) who defined the Permissible Exposure Levels (PELs) for personnel working in process industries. TLVs are of three types which are:

TLV-TWA (Time Weighted Average):
This is an average of normal working per day (8 hours) or 40 work hours per week on an average for which workers will be exposed.
TLV-STEL (Short-term Exposure Limit):
It is the maximum concentration to which workers can be exposed to for a period of up to 15 minutes continuously without suffering. The consequences are intolerable irritation, chronic or irreversible tissue changes, and narcosis of sufficient degree that reduces worker's efficiency considerably.
TLV-C (Ceiling Limit):

It is the concentration that should not be exceeded, even instantaneously or else it results in fatal death. The conversion of TLVs from ppm to mg/m³ is based on 760 mm Hg pressure at 25°C and a molar volume of 24.45 l, which is given by:

$$\text{TLV in mg/m}^3 = \frac{(\text{TLV in ppm}) \times (\text{gram molecular weight of substance})}{24.45}$$

$$(3.43)$$

$$\text{TLV in ppm} = \frac{(\text{TLV in mg / m}^3) \times 24.45}{(\text{gram molecular weight of substance})} \qquad (3.44)$$

3.24 Industrial Hygiene

Industrial hygiene is a science devoted to the identification, evaluation, and control of occupational conditions that cause sickness and injury. Identification is the determination of presence or possibility of workplace exposures. It requires a thorough study of chemical processes, operating conditions, and operating procedures (Michailidou et al., 2012). Sources of information for identifying the occupational hazards are through process design descriptions, operating instructions, safety reviews, and Material Safety Data Sheets (MSDS) (Engelhard et al., 1994). Evaluation is the determination of magnitude of exposure and control is the application of appropriate technology to reduce workplace exposures to the acceptable levels (Nivolianitou et al., 2006).

Science devoted to identification, evaluation, and control of occupational conditions that cause sickness and injury is called industrial hygiene. Identification is the determination of presence or possibility of workplace exposures. It requires a thorough study of chemical process, operating conditions, and procedures. The sources of information are: (i) process and design descriptions; (ii) operating instructions; (iii) safety reviews; and (iv) Material Safety Data Sheets (MSDS).

Evaluation is the determination of magnitude of exposure. The aim is to determine the extent and degree of employee's exposure to the physical and chemical hazards. Physical hazards are evaluated by comparing the existing strength with that of the threshold value, while chemical hazards are evaluated by comparing the concentration of toxicants with the allowable limit. Control is done by the application of appropriate technology to reduce the workplace exposures to the acceptable levels.

3.25 Exposure Evaluation: Chemical Hazard

3.25.1 Time Weighted Average Method

The concentration $C(t)$ in ppm or mg/m^3 of the chemical in air for worker shift time (t_w in hours) is given as:

$$C_{\text{TWA}} = \frac{1}{8} \int_0^{t_w} C(t)\, dt \tag{3.45}$$

For discrete average concentration C_i over a period of time T_i, TWA concentration is given by:

$$C_{\text{TWA}} = \frac{C_1 T_1 + C_2 T_2 + \cdots + C_n T_n}{8} \tag{3.46}$$

3.25.2 Overexposure at Workplace

The workplace is declared to be overexposed if $R > 1$. Overexposure limit is given by:

$$R = \frac{C_{\text{TWA}}}{\text{TLV}} \tag{3.47}$$

3.25.3 TLV–TWA Mix

For more than one chemical as a part of the inventory present in the plant, combined exposure from multiple toxicants should be calculated by:

$$C_{(\text{TLV–TWA})\text{mix}} = \frac{\displaystyle\sum_{i=1}^{n} C_i}{\displaystyle\sum_{i=1}^{n} \frac{C_i}{(TLV - TWA)_i}} \tag{3.48}$$

where, n is the total number of toxicants, C_i is the concentration of chemical (i) with respect to the other toxicants and $(\text{TLV-TWA})_i$ is for respective chemical (i). Workplace overexposure for mixture of multiple toxicants is given by:

$$R = \sum_{i=1}^{n} \frac{C_i}{(\text{TLV-TWA})_i} > 1 \tag{3.49}$$

3.26 Exposure Evaluation: Physical Hazards

Noise problems are common in process industries. Exposure to noise is measured in decibels. Decibel (dB) is a relative algorithm scale used to compare the intensities of two sounds and is given by:

$$\text{Noise intensity(dB)} = -10 \log\left[\frac{I}{I_0}\right] \tag{3.50}$$

where, I is the concerned sound intensity and I_0 is the reference sound intensity.

3.27 Industrial Hygiene Control

There are two major techniques for controlling industrial hygiene: (i) environmental control; and (ii) personal protection.

3.27.1 Environmental Control

The main aim is to reduce the concentration of exposed toxicants in the workplace. Good local ventilation should be provided to contain the hazardous substances. Dilution ventilation should be provided to control low-level toxics. Large openings that enable free passage of air will be useful in diluting the chemical concentration in the workplace. Use of wet methods minimizes contamination with dust. Good housekeeping will keep the toxicants and dusts contained within the workplace to a larger extent.

3.27.2 Personal Protection

The aim is to prevent or reduce exposure by providing a barrier between the worker and the workplace. Some of the safety measures that have to be used for protection are hard hat, safety glasses, chemical splash goggles, splash suit, ear plugs, etc.

3.28 Ventilation Hoods to Reduce Hazards

Most of the hoods that are provided in process industries assume plug flow as shown in Figure 3.19. Volumetric flow rate Q_v is given by:

$$Q_v = L \times W \times u \tag{3.51}$$

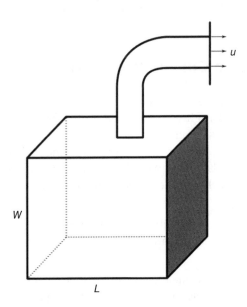

Figure 3.19 Plug flow through hood

Table 3.6   Non-ideal mixing factor, K for various ventilation conditions

Vapor concentration	Dust concentration	Mixing factor: ventilation conditions			
Parts per million (ppm)	Million particles per cubic feet (mppcf)	Poor	Average	Good	Excellent
Over 500	50	1/7	1/4	1/3	1/2
101–500	20	1/8	1/5	1/4	1/3
0–100	5	1/11	1/8	1/7	1/6

Table 3.7   Common types of chemical process accidents

Accident type	Chances of occurrence	Fatality chances	Chances of financial loss
Fire	High		Intermediate
Explosion	Intermediate		High
Toxic release	Low		

where, (L, W) are length and width of the hood and u is the required control velocity.

The non-ideal mixing factor, K, for various ventilation conditions are given in Table 3.6.

3.29 Elements to Control Process Accidents

Chemical plants are potential source of accidents. Oil spills, fire, explosions, reactor runaway and fugitive emissions are the potential hazards that are responsible for process accidents. Mechanical hazards include improper maintenance, tripping, falling, or moving equipment. These are main contributors to cause spills, fire, and explosions in chemical industries. Some of the common types of chemical accidents are given in Table 3.7.

Hazardous materials
Flammable materials, combustible materials, toxic chemicals, unstable materials, highly reactive reactants are called as hazardous materials.

Initiating events
Equipment malfunctions, containment failures, thermal runaway, human error in operations, maintenance, etc. are examples of initiating events in any accident.

Intermediate propagating events
Process parameters like pressure, temperature, flow rate and their deviations, toxic materials, reactive materials, ignition/explosion are examples of intermediate propagating events.

Intermediate mitigating events
Safety system responses (e.g., relief valves, grounding, back-up utilities), mitigation system responses (e.g., vents, blow-out walls/ceilings, containment dikes, flares, sprinklers, etc.), contingency operations like alarms, emergency procedures, personnel safety equipment, evacuations, security are different intermediate mitigating events that can control process accidents.

Accident consequences
Fire, explosions, dispersion of toxic chemicals are the severe consequences that arise from process accidents.

- *Steps in Risk analysis for process accidents*
- Predict the accident occurrence and its damage potential
- Reduce the envisaged risk well in advance of an accident
- Ensure system safety during operations

3.30 Methods for Chemical Risk Analysis

3.30.1 Qualitative Risk Analysis

This method is useful in predicting the undesired situations that may arise in any process system. This identifies the potential, chemical and mechanical hazards that result from the process industries. Some of the qualitative risk analysis methods are HAZOP study, PHA, FMEA, etc.

3.30.2 Quantitative Risk Analysis

This method is useful to evaluate the likelihood of occurrence of the accidents. This approach will be useful in determining the specific causes and consequences of the potential hazards along with their probable consequences. After identifying the specific causes and consequences, effectiveness of control measures and design modifications in the process system are evaluated. This method uses probability theory (PRA) such as FTA and ETA to analyze the causes and consequences of the accident.

3.31 Safety Review

There are two types of safety reviews: (i) informal safety review; and (ii) formal safety review. Informal safety review is generally applicable to the process where small changes are made to the existing process layout. This is generally applicable to small bench-scale labs. Formal safety review is used for analyzing new process plants or existing process plants where major changes are made in the functional layout. A team of experts need to be formed to develop and review the report and inspect the process plant.

3.32 Process Hazards Checklists

Process hazards checklists are the list of all possible hazards that the reviewer needs to consider. Checklists contain probable consequences of the identified hazards. It is a set of probable hazards that may arise and hence all the enlisted hazards may or may not apply to the present scenario. Nevertheless, checklists remind the areas of concern to the reviewers and stimulates them to revisit the recommendations/report summary to avoid even a remote risk occurrence. This can be applied in different stages: (i) during design conceptualization; (ii) during pilot plant operation; (iii) detailed design stage; (iv) routine checking stages; and (v) system design modification/expansion/decommissioning stages. For example, an automobile maintenance checklist could be the one that should be reviewed before driving for a vacation.

3.33 Hazard Surveys

It is a technique to identify and rank the hazards quantitatively (IS 1656-2000). This method is very simple if the method involves only the survey inventory of hazardous materials in a given facility. For example, Dow Fire and Explosion Index (F&EI), Dow-Chemical Exposure Index (CEI), etc. CEI is one of the most popular methods to study the toxic release and dispersion modeling that arise from process industries (Henselwood and Phillips, 2006).

3.34 Emergency Response Planning Guidelines

Emergency Response Planning Guidelines (ERPG) are published by American Industrial Hygienist's Association (AIHA). ERPG values are estimates of different concentration ranges where one might observe/experience any adverse effects. It is useful to identify the priority concerns in a process plant. It evaluates adequacy of the containment and identifies down-wind areas that will be affected by the release of toxic chemicals. These guidelines are helpful in developing community emergency response plans. There are three levels of ERPG:

ERPG-1:
It is the maximum airborne concentration below which all individuals could be exposed for a maximum period of 1 hour. During such exposures, they will experience only mild health effects; even objectionable odor should not be experienced.

ERPG-2:
It is the maximum airborne concentration below which all individuals could be exposed for a maximum period of 1 hour. During this period of exposure, they will experience health effects that are not irreversible or serious in nature.

ERPG-3:
It is the maximum airborne concentration below which all individuals could be exposed for a maximum period of 1 hour. During this period, they will experience adverse health effects but not of any life-threatening sort.

3.35 Chemical Exposure Index

Chemical Exposure Index (CEI) is a simple method of rating the relative health hazard to people residing in the neighborhood of chemical/process industries. CEI gives unit less risk index value, which is relative to various safety and environmental characteristics. It can be used for risk ranking of various options of safety aspects. This index value is used along with a decision analysis tool to check the process options and to meet the priorities of the process and safety as well.

To carry out CEI, the plan of the process plant and details of surrounding area, flow sheet, major piping, containment vessels, and chemical inventories are required as vital input. Physical and chemical properties of the material being investigated and their ERPG values are also required. A flowchart for estimating CEI is given in Figure 3.20.

Airborne quantity of the chemical released in air is given by:

$$AQ = 4.751 \times 10^{-6} D^2 P_a \left\{ \frac{MW}{T+273} \right\}^{1/2} \tag{3.52}$$

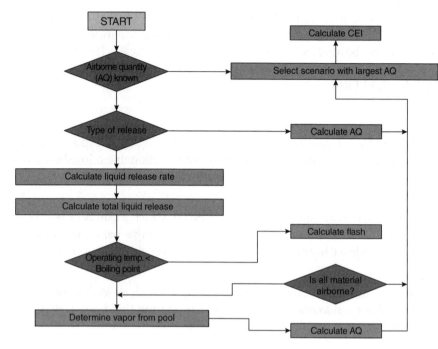

Figure 3.20   Steps for calculating CEI

where, AQ is the airborne quantity in kg/s, P_a is the absolute pressure in kPa gauge, MW is the molecular weight of the hazardous material being released in air, T is the process temperature in degree Celsius, and D is the hole diameter. For liquid release, liquid flow rate and the total liquid released are given by:

$$L = 9.44 \times 10^{-7} D_2 \rho_1 \sqrt{\left(\frac{1000 P_g}{\rho_1} + 9.8 \Delta h\right)} \tag{3.53}$$

$$W_T = 900 \times L \tag{3.54}$$

Discharge of some fuel in air will cause flashing. Occurrence of flashing is determined by comparing the operating temperature of the liquid to its normal boiling point. If the operating temperature is lower than its boiling point, flash fraction is considered to be nil. Alternatively, if the operating temperature is greater than its boiling point, vapor flash fraction (F_v) is computed using the following relationship:

$$F_v = \frac{C_p}{H_v}(T_s - T_b) \tag{3.55}$$

where, T_b is the normal boiling point of liquid (°C), T_s is the operating temperature of the liquid (°C), C_p is the average heat capacity of the liquid (J/kg/°C), and H_v is the heat of vaporization of the liquid (J/kg). If the vapor flash fraction computed from the above relationship is lesser than 0.2, then the AQ is calculated by the following relationship:

$$AQ_f = 5F_v \times L \tag{3.56}$$

If the flash fraction is greater than 0.2, then no pool is formed. In such cases, flash fraction is considered to be equal to liquid flow rate that is estimated using Equation (3.53). If the chemical released is a nonstandard mixture whose heat capacity is not known, then C_p/H_v can be taken as 0.0044.

In case flashing occurs, some liquid will be entrained as droplets. Some of these droplets will be quite small enough, which can travel with the vapor. Large droplets will fall into the ground to form a pool. If 20% of the released material flashes, then the entire stream is considered to be airborne; in that case, no pool is formed. Alternatively, if pool boiling takes place, then the AQ can be calculated using the following relationship:

$$AQ_p = 9 \times 10^{-4} \left(A_p\right)^{0.95} \frac{MW \times P_v}{T + 273} \tag{3.57}$$

where, P_v is the vapor pressure of the liquid at pool temperature in degree Celsius, T is the characteristic pool temperature in degree Celsius°C), MW is the molecular weight, and A_p is the pool area in m². The pool area is given by:

$$AQ_p = \frac{100W_p}{\rho_1} \tag{3.58}$$

where, W_p is the total mass entering the pool in kg, which is given by:

$$W_p = W_T \left(1 - 5F_v\right) \tag{3.59}$$

where, ρ_1 is the density (kg/m^3), W_T is the total liquid released, and F_v is the fraction released. AQ will be the sum of AQ of vapor fraction and the quantity of pool boiling, which is given by:

$$AQ = AQ_f + AQ_P \tag{3.60}$$

If AQ as computed earlier is found to be greater than the liquid flow rate computed earlier, then it is assumed that the AQ will be equal to the liquid flow rate. CEI is given by the following relationship:

$$CEI = 655.1\left[\frac{AQ}{ERPG-2}\right]^{1/2} \tag{3.61}$$

If the CEI value computed from the above equation is greater than 1000, then the CEI value is set to 1000. CEI calculations assume down-wind speed of $5\,m/s$ and normal weather conditions. Subsequently, hazard distances for different ERPG concentrations are computed using the following relationship:

$$HD = 6551\times\left[\frac{AQ}{ERPG}\right]^{1/2} \tag{3.62}$$

If the hazard distance exceeds $10\,000\,m$, then it is set to assume a maximum value of $10\,000\,m$. The CEI summary sheet is given in Figure 3.21.

After preparing the CEI summary, mitigation checklist is prepared as shown in Figure 3.22.

When established ERPG value of the mixture being released does not exist in the standard literature, then ERPG values are substituted as discussed next:

If ERPG-2 values does not exist, then one can use the workplace exposure guidelines as recommended by the DOW IGH, ACGIH TLV, or AIHA WEEL. Alternatively, STEL or ceiling value TWA can be used in place of the ERPG-2 value. If ERPG-3 values does not exist, then one can assume this value as five times of that of the value of ERPG-2. If EPRG-1 does not exist, then one can either consider the odor threshold value or assume one-tenth of that of the value of ERPG-2.

Now let us consider an example of a chemical process shown in Figure 3.23 for calculating the CEI. The purpose is to determine which process piping or equipment has the greatest potential for the release of significant quantities of acutely toxic chemicals. Evaluating several scenarios will aid in determining the largest potential airborne release. Process conditions such as temperature, pressure, physical state, and pipe size should be considered since they have a significant impact on airborne release rates.

Chemical Exposure Index Summary

Plant: _____   Location: _____

Chemical: _____   Total quantity in the plant: ___

Largest single containment: _____

Pressure of the containment: _____

     Temperature: _____

1. Scenario being evaluated: _____
2. Airborne release rate from scenario: _____ kg/s
3. Chemical exposure index: _____
4. Concentration of the chemical

	Concentration	Hazard distance
ERPG-1	_____	_____
ERPG-2	_____	_____
ERPG-3	_____	_____

5. Distance to

    Public property line: _____

    Other in-house facility: _____

    Non-company plant or business: _____

6. CEI computed above and the hazard distances establish the level of review needed
Further review required: YES/NO
7. If further review is required, prepare a review package
8. List any sights, odors, or sounds that might arise from the facility being inspected, which might cause public concerns or injuries (e.g., smoke, larger relief valves, odors below hazard levels but still objectionable etc.)

_____

_____

_____

Prepared by: _____

Reviewed by: _____

Plant Superintendent or Manager:
_____dated: _____

Site review representative:
_____dated: _____

Additional management of review (if required):
_____dated: _____

Figure 3.21 Chemical exposure index summary

Complete (✓)	Risk Reducing Factors
_____	1. All pressure vessel and relief device systems properly registered and inspection up to date and documentation complete. (No expansion joints or glass devices.)
_____	2. All hoses inspected and tested regularly.
_____	3. All operational controls and systems designed and routinely tested to "fail-safe."
_____	4. Critical Instrument Program up to date (e.g., redundant high level and temperature alarms, shutdowns, etc.)
_____	5. Operating Discipline complete and up to date.
_____	6. Vapor Detectors properly placed and tested regularly.
_____	7. Appropriate engineering specifications properly applied (e.g., lethal service, welded fittings, etc.)
_____	8. Are relief vents on toxic containers designed to minimize atmospheric emissions? How? (circle) Scrubber, Flare or _____
_____	9. Failure analysis and nondestructive testing carried out where needed (e.g., X-ray, vibration analysis or monitoring, acoustical emission, piping flexibility – hot and cold).
_____	10. Physical barriers in place (for traffic, cranes, etc.)
_____	11. Designed for excess pressure, if needed (e.g., pipelines in certain areas, tank cars, trucks, etc.).
_____	12. All personnel properly trained to understand hazards and emergency responses.
_____	13. Emergency Procedures (relating to this exposure potential) in place and annual drill held.
_____	14. Safety Rules and Safety Standards regularly reviewed and enforced.
_____	15. Loss Prevention Principles and Minimum Requirements appropriately applied.
_____	16. Technology Canter Guidelines appropriately incorporated.
_____	17. Reactive Chemical Review complete and up to date.
_____	18. Loss Prevention Audit complete and up to date.
_____	19. Technology Center Audit complete and up to date.
_____	20. All new operations and modifications underwent safety pre-startup audit.
_____	21. Management of Change procedures written and utilized.

Figure 3.22 Mitigation check list

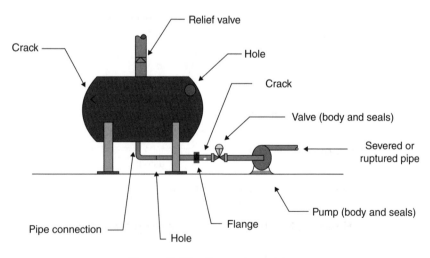

Figure 3.23 Example problem

The scenario selection for CEI calculation is the process pipeline. The scenarios of the process pipeline failure are shown in Figure 3.24. Rupture of the largest diameter process pipe is discussed. For smaller than 50 mm diameter, full bore rupture is considered. For 50–100 mm diameter, rupture equal to that of 50 mm diameter pipe is being considered. For greater than 100 mm diameter, rupture area equal to 20% of that of the pipe cross-section area is considered. Different failure scenarios are given in Figure 3.24.

Releases from all scenarios are assumed to continue for a duration of at least 5 minutes (Amendola et al., 1992).

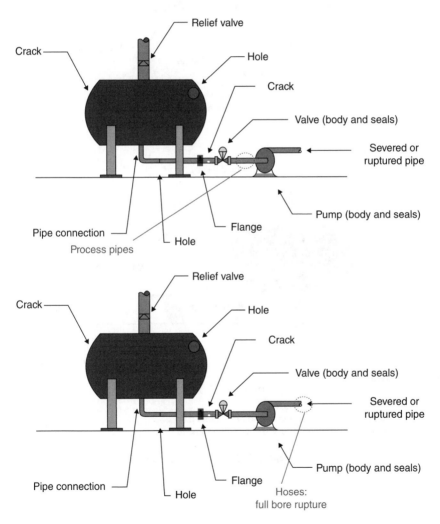

Figure 3.24 Release scenarios

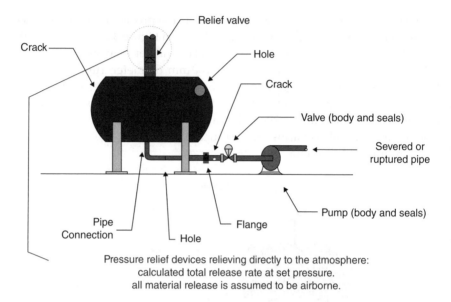

Pressure relief devices relieving directly to the atmosphere:
calculated total release rate at set pressure.
all material release is assumed to be airborne.

Vessels:
rupture based on largest diameter process pipe
Attached to the vessel using pipe criteria above.

Figure 3.24   (*Continued*)

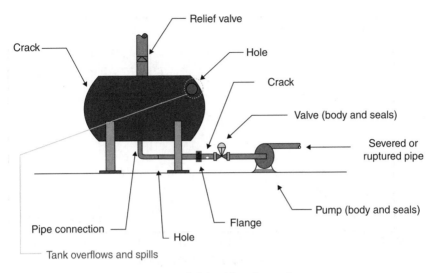

Figure 3.24 (*Continued*)

3.36 Guidelines for Estimating Amount of Material Becoming Airborne Following a Release

AQ refers to the total quantity of material entering the atmosphere over time. This can be a direct vapor release or that arises from the liquid flashing or pool evaporation (Chandrasekaran, 2010a). CEI scenarios consider material to be released both as liquid or vapor; variation of CEI with AQ is given in Figure 3.25 (TNO, 1999).

3.36.1 Example Problem on Ammonia Release

Ammonia is stored in a 12-ft diameter by 72-ft long horizontal vessel under its own vapor pressure at ambient temperature (30°C or 86°F). The largest liquid line out of the vessel is 2- inch diameter (50.8 mm).

P_g (pressure inside vessel) = 1064 kPa gauge; T (temperature inside vessel) = 30°C; T_b (normal boiling point) = −33.4°C; ρ_1 (liquid density) = 594.5 kg/m³; C_p/H_v = 4.01 × 10⁻³; Δh (height of liquid in tank) = 3.66 m; D (diameter of hole) = 50.8 mm; MW (molecular weight) = 17.03

Solution:
Step 1: Estimate liquid released

$$L = 9.44 \times 10^{-7} D^2 \rho_1 \sqrt{\left(\frac{1000 P_g}{\rho_1} + 9.8 \Delta h\right)} \, \text{kg/s}$$

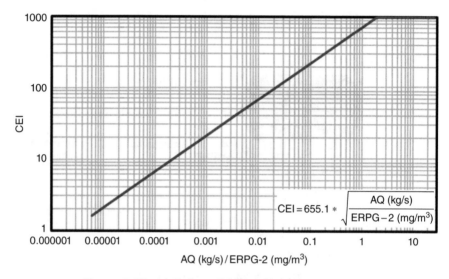

Figure 3.25 Variation of CEI with airborne quantity

Substituting the values, we get:

$$L = 9.44 \times 10^{-7} (50.8)^2 (594.5) \sqrt{\left(\frac{1000(1064)}{594.5} + 9.8(3.66) \right)}$$

$$L = 61.9 \text{kg/s}$$

Step 2: Estimate flash fraction
$F_v = C_p / H_v (T - T_b)$
$F_v = 0.00401(30 - (-33.4))$
$F_v = 0.254$
Since $F_v > 0.2 \rightarrow AQ = L$
$AQ = 61.9 \text{kg/s}$

Step 3: Calculate the CEI
Where ERPG-2 = 139 mg/m³
CEI = 655.1(AQ/ERPG-2)1/2
CEI = 655.1(61.9/139)1/2
CEI = 437

Step 4: Calculate hazard distances
For ERPG-2 = 139 mg/m³
HD = 6551(AQ/ERPG)1/2
HD = 6551(61.9/139)1/2
HD = 4372 m
For ERPG-1 = 17 mg/m³
HD = 6551(AQ/ERPG)1/2
HD = 6551(61.9/17)1/2
HD = 12 500 m
For ERPG-3 = 696 mg/m³
HD = 6551(AQ/ERPG)1/2
HD = 6551(61.9/696)1/2
HD = 1953 m

3.36.2 Example Problem on Chlorine Release

The ¾-inch vapor connection on a 1-ton chlorine cylinder stored at ambient temperature (30°C or 86°F) has fractured. The chlorine storage vessel is shown in Figure 3.26.

P_g (pressure inside cylinder) = 788.1 kPa; P_a (absolute pressure) = 889.5 kPa; MW (molecular weight) = 70.91; T (storage temperature) = 30°C; D (diameter of hole) = 19 mm.

Solution:

Step 1: Determine AQ
$AQ = 4.751 \times 10^{-6} \, D^2 \, P_a (MW/(T + 273))^{1/2}$
$AQ = 4.751 \times 10^{-6} \, (19)^2 (889.5)(70.91/(30 + 273))^{1/2}$
$AQ = 0.74 \, kg/s$

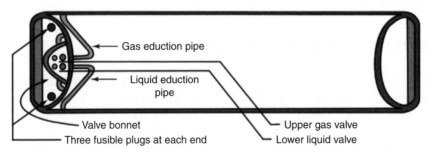

Figure 3.26 Chlorine storage vessel

Step 2: Calculate the CEI
Where ERPG-2 = $9\,\text{mg}/\text{m}^3$
$\text{CEI} = 655.1(\text{AQ}/\text{ERPG-2})^{1/2}$
$\text{CEI} = 655.1(0.74/9.0)^{1/2}$
$\text{CEI} = 188$

Step 3: Calculate hazard distances
For ERPG-2 = $9\,\text{mg}/\text{m}^3$
$\text{HD} = 6551(\text{AQ}/\text{ERPG})^{1/2}$
$\text{HD} = 6551(0.74/9)^{1/2}$
$\text{HD} = 1878\,\text{m}$
For ERPG-1 = $3\,\text{mg}/\text{m}^3$
$\text{HD} = 6551(\text{AQ}/\text{ERPG})^{1/2}$
$\text{HD} = 6551(0.74/3)^{1/2}$
$\text{HD} = 3254\,\text{m}$
For ERPG-3 = $58\,\text{mg}/\text{m}^3$
$\text{HD} = 6551(\text{AQ}/\text{ERPG})^{1/2}$
$\text{HD} = 6551(0.74/58)^{1/2}$
$\text{HD} = 740\,\text{m}$

3.37 Quantified Risk Assessment

Risk analysis can be defined as systematic identification and description of risk to personnel, environment, and equipments. Quantified Risk Assessment (QRA) therefore has to be focused on identification of applicable hazards and description of applicable risks to personnel, environment, and assets. The analytical elements of risk assessment include identifying the relevant hazards and to assess the risks arising from them. These also include identification of initiating events, causes, and consequences.

3.38 Hazard Identification (HAZID)

The identification of initiating events are generally called as HAZID. A broad review of possible hazards and sources of accidents are done initially. Critical hazards are then classified for the subsequent level of analyses. Some of the hazard identification tools employed in the industries are checklists, accident and failure statistics, HAZOP, comparison with detailed studies, and experience from previous similar projects, concepts, systems, equipment, and operations.

3.39 Cause Analysis

The cause analysis identifies the causes that may lead to initiating events and assess probability of such events. It is important to identify the cause of hazards or initiating events, as they will be the starting point of potential accident sequences. Identifying the initiating events is done through a qualitative approach and probability of initiating events is done through quantitative approach. The qualitative analysis identifies the causes that lead to the initiating events and possible combinations that may result in the occurrence of incidents. Some of the qualitative analysis techniques that are commonly used in oil and gas industries are HAZOP, Preliminary Hazard Analysis (PHA), Failure Mode and Effect Analysis (FMEA), and human error analysis such as task analysis and error mode analysis. Some of the quantitative techniques are Fault Tree Analysis (FTA), Event Tree Analysis (ETA), synthesis models, Monte Carlo simulation, BORA methodology, etc.

3.40 Fault Tree Analysis (FTA)

FTA links the final event with the basic events through a sequence of intermediate events. This is a deductive logic in which it progresses backward starting from the final event. The final event may be fire or an explosion. Events are divided into final, intermediate, and basic events, which are consecutively interlinked through different gates as shown in Figure 3.27.

The probability of final event is calculated using AND gate and OR gates.

Example problem

Consider an example to construct a fault tree for the formation of a fireball as a final event. Chances of LPG release from the pressurized vessel are due to weld failure or opening of relief valve by the operator (Johnson and Cornwell, 2007; Pontiggia et al., 2011). Flammable mixture thus formed will get ignited if there is any ignition source and leads to fireball. The events demonstrated using FTA as shown in Figure 3.28. The final event due to LPG release is fireball. From the fault tree, it can estimated as follows:

Probability of LPG release	=	$P_1 + P_2$
Probability of ignition	=	$P_3 + P_4$
Probability of fireball, P_f	=	$(P_1 + P_2) \times (P_3 + P_4)$

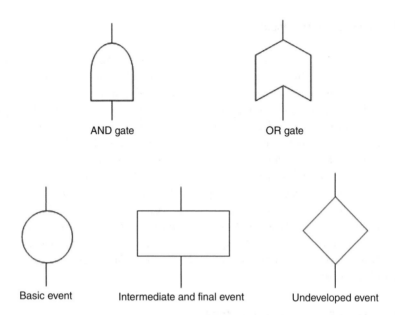

Figure 3.27 Logic gates

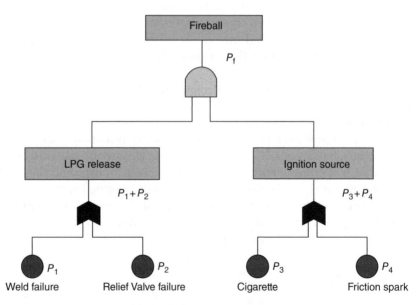

Figure 3.28 LPG release

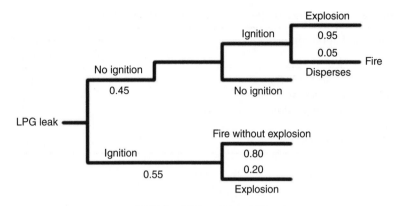

Figure 3.29 Event tree

3.41 Event Tree Analysis (ETA)

ETA is an inductive approach in which it progresses forward through subsequent events, leading to consequences. The ETA is illustrated through the following example involving the leakage of LPG from a road tanker from highway tunnel. The probability of encountering an ignition source is 0.45. A fire, if formed, could transmit to an explosion with a probability of 0.20. The event tree is given in Figure 3.29.

3.42 Disadvantages of QRA

Different approaches in QRA give different results. Scenario selection depends on the expertise of the risk assessor. Change in environmental conditions, that is, temperature, humidity, and wind speed can alter the results significantly. Each software model simulates different results for the same release scenarios. All countries do not have statutes specifying the acceptable risk limits. Database used for probability can also make a significant difference in the results.

3.43 Risk Acceptance Criteria

The risk acceptance criterion is important to compare the calculated risk of the plant with that of the acceptable limits. Different countries follow different acceptance criteria for the process industries. A brief summary of these risk criteria are shown in Table 3.8.

Table 3.8 Risk criteria in various countries (IS15656:2006)

Authority and application	Maximum tolerable risk (per year)	Negligible risk (per year)
VROM, The Netherlands (new)	1E–6	1E–8
VROM, The Netherlands (existing)	1E–5	1E–8
HSE, UK (existing hazardous industry)	1E–4	1E–6
HSE, UK (nuclear power station)	1E–5	1E–6

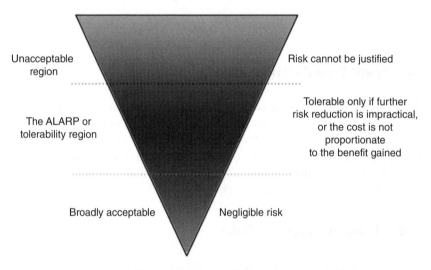

Figure 3.30 ALARP triangle

The risk acceptance criteria as per HSE, UK (for existing hazardous industry) can be expressed in form of ALARP (As Low As Reasonably Practical) triangle shown in Figure 3.30. From the figure it can be seen that the negligible risk or minimum acceptance criteria of risk is 1E-6 per year. Below this line the risk is broadly acceptable and risk is considered to be negligible. The maximum acceptable risk is 1E-4 per year. Above this risk, the risk is intolerable or unacceptable. In between is the ALARP region, in which the risk is tolerable only if further risk reduction is impractical, or the cost is not proportionate to the benefit gained.

3.44 Hazard Assessment

Hazard assessment is carried out to prevent the work-related injury or illness to the workers. Assessing hazards deal with a careful insight of situations that could go wrong for the workers and assets in the plant. Hazard assessment is done through a series of questions that consists of what if the system fails, what if the safety valve is not working, etc.

(a) When?
   Hazard assessment is done in the plant when a new work process is introduced or if there is any change in the work process or operation. Before the addition or installation of new critical equipments or the segment of the plant that has unhealthy working conditions, it is mandatory to carry out hazard assessment.

(b) How?
   The hazard assessment is done through the procedures explained in the following section. All types of work and their related activities need to be identified and listed. While hazard assessment starts from this step, identifying hazardous activities among a list of operations being carried out in a process plant needs sound backing of experience.

3.45 Identify Hazards

Identifying hazards for each of the work-related activities is carried out by four methods: physical inspection, task or job hazard analysis, process analysis, and incident investigation findings.

3.45.1 Prioritizing Hazards

The high hazardous work activities that are identified need to be also prioritized. This includes preparation of incident reports, listing the severity of the incidents, data collection based on the discussion with the personnel on board, details that are borrowed from the material data sheet, data based on the incident's statistics and reports and safety audits.

Incident reports
In this section, work activities that are resulted in near-miss incidents are reported and documented.

Severity of these incidents
The work activities that are resulted in serious injury or fatality with higher intensity are noted down.

Talk to workers
Workers may be aware of the unreported near-miss incidents, which can be a vital input for hazard analysis.

Material safety data sheet
It will be helpful in identifying the unrecognized or underestimated hazards involving substances.

Industry incident statistics and reports
The incidents that occurred in the similar type of industries are identified. This information could give a better idea of extrapolating the possible hazardous scenarios in the present case.

Inspection reports and safety audits
This will be helpful in gaining knowledge about existing potential problems.

3.46 Risk Assessment

Risk assessment gives a better picture of various scenarios of the plant as this also includes the consequences along with the probability of occurrence of the incidents. It is also helpful in prioritizing the risk from higher to the lower. Focus can be directed toward the events with higher risk envisaged. Risk assessment includes details about severity of the accident, probability of occurrence of the accident, and its frequency.

3.46.1 Identify and Implement Hazard Controls

Hazard control measures that already exist in the plant are identified. Additional safety measures are recommended for if the assessed risk is unacceptable or if the risk is not within the permissible limits.

3.46.2 Communicate

Information about the risk ande safety measures needs to be communicated among the workers as they have the highest potential of risk that arise from foreseen hazards. Proper training needs to be imparted about the nature of hazards, their probability of occurrence, cost implications, possible controls to mitigate the consequences and their responsibilities in risk management.

3.47 Evaluate Effectiveness

Effectiveness of control measures must be checked during inspection or working hours of the plant. Effectiveness of various control mechanisms that exist in the plant can be evaluated by asking questions to the workers, such as: (i) have the control measures solved the problem?; (ii) did the control measures create any new hazards?; (iii) are any new control measures required as recommended by previous inspection reports?; and (iv) are control measures updated based on the incident reports being analyzed? Hazards can be easily identified based on the answers to such questions. Subsequently, the identified hazards are prioritized according to the critical hazard points. Hazard controls are then implemented to prevent major accidents. Effectiveness of these control measures has to be evaluated for better safe working environment in periodic intervals, through inspection and safety audits.

3.48 Fatality Risk Assessment

Fatality risk assessment is one of the important elements of a quantified risk assessment. It involves a lot of uncertainties due to insufficient data available for the analysis. Therefore modeling fatalities is very complex. Ratio of fatalities to injuries in the exploration and production is about 1 : 1400 during the past 10 years. It is much clear that the injury statistics are more than fatalities and therefore focus has to be on reducing the injuries, which in turn will reduce the fatalities. If the fatality risk is assessed and focus on risk-reducing measures, then probably the risk of having injuries also can be reduced significantly.

3.48.1 Statistical Analysis

Statistical analysis of fatality risk is used when there exists sufficient database of accidents. Uncertainties are less extensive in a statistical analysis. Therefore, calculation of fatality risk, based on statistical analysis is often used for occupational hazards.

3.48.2 Phenomena-Based Analysis

This type of analysis includes chain of events such as cause of fire, fire loads, responses, and effects on personnel from fire loads. This approach describes the behavior of persons during a major accident. These analyses include various steps that a person has to go through in order to save his or her life in a major accident. One of the main disadvantages of this type of analysis is that there are uncertainties involved in each step, which may lead to higher level of uncertainty in the whole analysis.

3.48.3 Averaging of FAR Values

FAR values are averaged over separate groups considered in the plant. Groups are categorized according to the departments in the plant, such as office, process, production, drilling, etc. FAR varies from one section to another, which implies that the personnel working in different sections will have different FAR levels.

3.49 Marine Systems Risk Modeling

In the recent past, there is a huge development in the offshore industry in which the surface installations have been taken over by offshore installations. Offshore installations mainly consist of floating or fixed installations, which envisage higher probability of failures. Common causes arise from ballast systems, anchoring systems, loss of buoyancy, etc.

3.49.1 Ballast System Failure

Loss of stability results from either a single point of failure or a combination of multiple failures. Some of the hazards of ballast failure are: (i) failure of pumps, valves, and control systems; (ii) operational failure; (iii) loss of weights due to anchor-line failure; (iv) ballast system failure during transition of mobile units; and (v) failure during operation of the loading system, which leads to abnormal weight condition.

3.50 Risk Picture: Definitions and Characteristics

According to the international standards, risk is defined as a combination of probability of events and their consequences. Risk is therefore a product of probability of occurrence of an undesirable event (realization of hazard) and the corresponding consequence. For a series of events that are responsible for risk, it can be expressed as a sum of the product of their probability of occurrence and consequences. This is given by:

$$R = \sum_i \left(p_i \cdot C_i \right) \tag{3.63}$$

where, p is the probability of accidents, C is the consequence of accidents. In general, in offshore industries, probability of occurrence of hazardous events is very low, but consequences are very high as they result in catastrophic accidents; this amounts to a high risk picture in offshore industries. This also necessitates to understand the fact that risk is more toward the loss and not toward the precaution of occurrence. It means that the financial component of risk, which leads to high economic loss is of major concern in offshore industries. While risk can be classified as personnel, environmental, and asset risk, personnel risk can be further subdivided into fatality risk and impairment risk. Asset risk can be subdivided into material damage risk and production delay risk.

3.51 Fatality Risk

Fatality risk is classified into platform fatality, individual, and societal risk.

3.51.1 Platform Fatality Risk

It includes the estimation of Potential Loss of Life (PLL), which is also known as fatalities per platform year. From the PLL, one can deduce the individual and group risk or societal risk. PLL is give by:

$$PLL = \sum_n \sum_j \left(f_{nj} \cdot c_{nj} \right) \tag{3.64}$$

where, f_{nj} is the annual frequency of accident scenario (n) with the personnel consequence (j), C_{nj} is the expected number of fatalities for the known accident scenario, n is the total number of accident scenarios in all event trees, and j is the total number of personnel consequences (e.g., immediate, escape, evacuation, and rescue fatalities).

3.51.2 Individual Risk

Individual risk can be defined as frequencies at which the individual may be expected to sustain a given level of harmfulness that arise from the realization of hazard. It is the ratio of number of fatalities to number of people (exposed) at risk. It is expressed in terms of number of fatalities per average year. Individual risk can be expressed both mathematically and graphically. While the following equation is used to express the individual risk mathematically, graphical representation is given in Figure 3.31.

$$FAR = \frac{PLL \cdot 10^8}{\text{Exposed hours}} \qquad (3.65)$$

3.52 Societal Risk

Societal risk is defined as the relationship between the frequency and number of people suffering a given level of harmfulness that arise from the realization of hazards. It is expressed in terms of FN curve, where F identifies the frequency of occurrence of events and N denotes the number of fatalities. A typical plot is shown in Figure 3.32.

Societal risk is calculated using Equation (3.66). $N_{edf|o}$ is the number of people exposed to the accidental consequences. The count will also consider the hazardous environmental conditions that are responsible for such accidents namely: (i) adverse weather conditions; (ii) wind speed and direction of flow; and (iii) type of accident (flammable, toxic, or explosion).

$$N_{edf|o} = \iint n_{x,y} P_{d,x,y|o} \, dx \, dy \qquad (3.66)$$

where, $n_{x,y}$ is the number of population per cell or grid considered for the calculation and $P_{d,x,y|o}$ is the probability of population in the grid.

3.53 Impairment Risk

Impairment risk is an indirect way of expressing risk that is crucial for personnel safety. Impairment frequencies are calculated for main safety functions on the basis of physical modeling of responses to the foreseen accident loading. Impairment risk is an assessment method that is based on consequences of an accident, not in terms of fatalities but impairment of safety functions. According to UK legislations, the impairment of safety functions include life support, command safety, and evacuation safety. The frequency of impairment is given by:

$$f_{imp,i} = \sum_n f_n \cdot P_{imp,n,i} \qquad (3.67)$$

where, $P_{imp,n,i}$ is the probability of impairment scenario (n), for safety the function and n is the total number of accident scenarios being considered in the model.

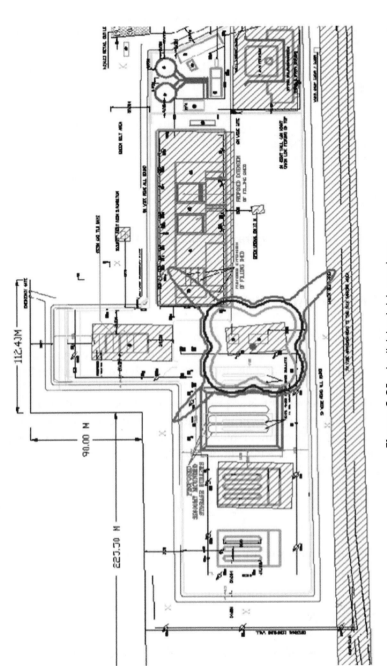

Figure 3.31 Individual risk contour

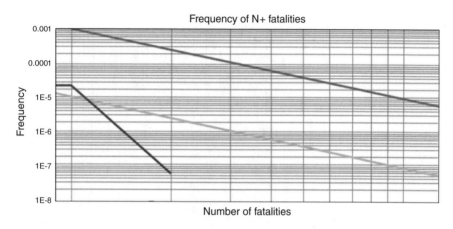

Figure 3.32 Societal risk

3.54 Environmental Risk

Environmental risk includes leaks from storage vessels, pipelines, or blowouts from offshore industries. Process pipeline leakages are one of the most common accidents in oil and gas industries, which can be controlled to prevent worse consequences (Rodante, 2004; Dziubinski et al., 2006). In such cases, risk to environment is expressed as the expected value of spilled amount or frequency of events with similar consequences; consequence is measured in restoration time, which is the time needed by the environment to recover after a spill (Webber et al., 1992). The expected value of spilled amount is given by:

$$Q = \sum_n f_n \cdot q_n \tag{3.68}$$

where, f_n is the frequency of accident for scenario (n) and q_n is the quantity of spill for each scenario.

3.55 Asset Risk

Asset risk includes risk to equipments and structures, which causes disruption of production. Risk is expressed in terms of material damage and production delay. Asset risk can be expressed as expected damage to the structures and equipment or expected duration of production delay. The expected value of damage per year, D is give by:

$$D = \sum_n f_n \cdot d_n \qquad (3.69)$$

where, d_n is the extent of damage or duration of delay for scenario (n) and f is the frequency of each such damage scenario.

3.56 Risk Assessment and Management

Risk is based on probability of occurrence and the severity of accidents. It depends on the hazards present in the industry. Hazard is a physical or a chemical condition that has the potential to cause damage to people and environment. Hazards lead to risks. Risk can be assessed either quantitatively or qualitatively. Risk can be quantitatively defined by the product of probability of occurrence of event and its severity. But the calculation of risk is very difficult and if the probability of event is known it is not necessary that the severity has to be known and vice versa. There are many factors that add to this complexity as safety, cost, schedule, technical, and geo-physical conditions. Most of the decisions include one or more of these to be considered. An important part of risk management is deciding which types of risk to assess and how they should be compared to take decisions.

3.57 Probabilistic Risk Assessment

Probabilistic Risk Assessment (PRA) is a quantitative procedure to measure technical risks. This procedure includes identifying hazards and their initiating events, identifying mitigating safety measures, tracing possible chains of events, quantifying all individual probabilities, and severities to calculate risk. The probabilistic risk analysis is done through reliability analysis. Reliability is preventing the system from failure; therefore it is related to risk. When the system becomes complex, it is very difficult to account for various combinations of failures. Therefore, methods like fault tree and event tree are developed to facilitate such analysis.

3.58 Risk Management

Managing risks frequently requires more than a series of calculations; in real time practice, it is rather very difficult to manage. By conducting PRA or other risk analysis method, risk involved in the process can be determined and can also be prioritized (Bonvicini et al., 1998). It is important for the oil and gas industry to understand how to manage such risks that arise from both the process and plant and plant design. Unfortunately for technical analysts, perceptions and decisions related to risk are very complex. They even appear irrational. It is therefore important to understand that risk is not quantified in absolute terms but on relative terms; this is done by prioritizing risks within various scenarios.

3.58.1 Risk Preference

Risk preference is simply an individual's feeling or opinion about risk. Like any preference, risk preference can range from desire to avoidance. The trouble is that most people actually seem to exhibit both the types of risk preference.

Exercises 3

1. According to UK Health and Safety Executive, Acceptable FAR is ....

   Fatality Accident Rate

2. The diffusion of toxicants through skin membrane is called ....

   Dermal absorption.

3. .... gives unit less risk index value which is relative to various safety and environmental characteristics.

   CEI

4. If the response of interest is death or lethality, the response versus log dose curve is called ....

   Lethal dose curve.

5. What is the acceptable limit for carcinogens according to US EPA criteria?
   (a) E–6   (b) E4   (c) E6   (d) E5

   E6 (c)

6. Entry of toxicants through skin membrane is called:
   (a) Dermal absorption   (b) Ingestion   (c) Inhalation   (d) Injection

   Dermal absorption (a)

7. If the response to the agent causes an undesirable response that is not lethal but is irreversible, for example, liver damage or lung damage, the response–log dose curve is called:
   (a) Lethal dose   (b) Effective dose   (c) Toxic dose   (d) Lethal concentration

   Toxic dose (c)

8. Concentration that should not be exceeded, even instantaneously is called:
   (a) TLV STEL   (b) TLV-C   (c) TLV-TWA   (d) ERPG

   TLV-C (b)

9. What are the various types of doses?

   Lethal dose, effective dose, toxic dose, lethal concentration.

10. Define toxic dose.

    TD (Toxic dose)—if the response to the agent is toxic (it causes an undesirable response that is not lethal but is irreversible, for example, liver damage or lung damage), the response–log dose curve is called TD curve.

11. .... represent conditions to which all workers will be repeatedly exposed every day without adverse health effects.

    TLV

12. .... is the maximum concentration to which workers can be exposed for a period of up to 15 minutes continuously without suffering.

    TLV– STEL

13. ....are estimates of concentration ranges where one might observe adverse effects.

    ERPG

14. .... is evaluated by comparing existing strength with threshold value.

    Physical hazard

15. What are the different types of TLV's?

    TLV-TWA, TLV-STEL, TLV-C

16. Define industrial hygiene. Explain the various steps involved.

Science devoted to identification, evaluation, and control of occupational conditions that cause sickness and injury.

Identification: Determination of presence or possibility of workplace exposures.
Evaluation: Determination of magnitude of exposure.
Control: Application of appropriate technology to reduce workplace exposures to acceptable levels.

17. A press cleaner is monitored for exposure to ethanol. The data are:

Time period (number)	Concentration (ppm)	Sample duration (h)
1	410	1.5
2	250	3.5
3	75	2

Calculate TWA.

$$TWA = \frac{410\,ppm \times 1.5\,h + 250\,ppm \times 3.5\,h + 75\,ppm \times 2\,h}{1.5\,h + 3.5\,h + 2\,h}$$

$$= \frac{1640\,ppm \cdot h}{7\,h} = 234\,ppm$$

$$8h - TWA = \frac{1640\,ppm \cdot h}{8\,h} = 205\,ppm$$

18. Trichloroethane (a solvent) has a biological half life of 16 hours. What TLV would be appropriate for people who work 3 shifts of 12-hour per week when they are likely to be exposed to the compound? TLV and PEL for the chemical is 10 ppm.

PEL: 6.7 ppm; TLV: 5 ppm

$$PEL = \frac{8\,h}{12\,h} \times 10\,ppm = 6.7\,ppm$$

$$TLV = \frac{8\,h}{12\,h} \times \frac{24 - 12\,h}{16\,h} \times 10 = 5\,ppm$$

19. Ammonia is stored in a 4-m diameter and 18-m long horizontal vessel under its own vapor pressure at temperature 42°C. The largest liquid line out of the vessel is 2-inch diameter (50.8 mm).

P_g (pressure inside vessel) = 1050 kPa gauge
T (temperature inside vessel) = 42°C

T_b (normal boiling point) $= -33.4°C$
ρ_1 (liquid density) $= 594.5\,\text{kg/m}^3$
$C_p/H_v = 4.01 \times 10^{-3}$
Δh (height of liquid in tank) $= 4\,\text{m}$
D (diameter of hole) $= 50.8\,\text{mm}$
MW (molecular weight) $= 17.03$
Calculate the hazard distances for all ERPG values. (ERPG-1:17 mg/m^3; ERPG-2: 139 mg/m^3; ERPG-3: 696 mg/m^3)

$$L = 9.44 \times 10^{-7} \times 50.8^2 \times 594.5 \times \sqrt{\frac{1000 \times 1050}{594.5}} + 9.8 \times 4 = 61.61\,\text{kg/s}$$

$$F_v = 4.01 \times 10^{-3} \left(42 - (-33.4)\right) = 0.3$$

Since $F_v > 0.2 \rightarrow AQ = L$
$AQ = 61.61\,\text{kg/s}$

CEI
  $CEI = 655.1(61.61/139)^{1/2} = 436.14$ for ERPG-2

Hazard Distances
  $HD = 6551(61.9/17)^{1/2} = \textbf{12471.21 m}$
  $HD = 6551(61.9/139)^{1/2} = \textbf{4361.4 m}$
  $HD = 6551(61.9/696)^{1/2} = \textbf{1949.08 m}$

20. One-inch vapor connection on a 1-ton chlorine cylinder stored at ambient temperature (30°C or 86°F) has fractured.
  P_g (pressure inside cylinder) $= 600\,\text{kPa}$
  P_a (absolute pressure) $= 889.5\,\text{kPa}$
  MW (molecular weight) $= 70.91$
  T (storage temperature) $= 40°C$
  D (diameter of hole) $= 25\,\text{mm}$.
  Calculate the hazard distances for all ERPG values. (ERPG-1: 3 mg/m^3; ERPG-2: 9 mg/m^3; ERPG-3: 58 mg/m^3)

Airborne Quantity (AQ)
  $AQ = 4.751 \times 10^{-6} (25)^2(889.5)(70.91/(40+273))^{1/2} = \textbf{1.26 kg/s}$

CEI
  $CEI = 655.1(1.26/9.0)^{1/2} = 245.12$ for ERPG-2

Hazard Distances
  $HD = 6551(1.26/3)^{1/2} = \textbf{4245.53 m}$
  $HD = 6551(1.26/9)^{1/2} = \textbf{2451.2 m}$
  $HD = 6551(1.26/58)^{1/2} = \textbf{965.56}$

21. ………….. identifies various hazards through a qualitative review of possible accidents that may occur.

    HAZOP

22. …………. is the minimum oxygen concentration below which combustion is not possible, with any fuel mixture.

    Limiting oxygen concentration (LOC)

23. ………….. is a rapid expansion of gases resulting in a rapidly moving pressure or shock waves.

    Explosion

24. What are the different published guidelines and standards for the requirements and recommended reporting format of failure mode and effects analyses?

    SAE J1739, AIAG FMEA-3, and MIL-STD-1629A.

25. Define risk as per (i) ISO 2002 (ii) ISO 13702

    ISO 2002: Combination of the probability or an event and its consequence.
    ISO 13702: A term which combines the chance that a specified hazardous event will occur and the severity of the consequences of the event.

26. A person is exposed to heat radiation of $15 \, kW/m^2$ for 1 minute. The dose as a function of intensity of heat radiation I and time t is given by $D = (I(W/m^2))^{1.33} \times t(s)$. The Probit for fatal injury is given by $Y = k_1 + k_2 \ln D$, where $k_1 = -37.23$ and $k_2 = 2.56$. Determine the probability of fatal injury. You can assume the relation between percent response R and Probit Y to be given by the linear equation $R = 38.2Y - 141$.

    $Y = k_1 + k_2 \ln D$
    $D = (15\,000)1.33 \times 60$
    $Y = -37.23 + 2.56 \ln [150\,001.33 \times 60] \, Y = 6$
    The value $Y = 6$ corresponds to $R = 38.2 \times 6 - 141 = 88$, suggesting the probability of fatal injury to be 88%

27. .... is formed due to the catastrophic failure of the pressurized vessel containing fluid above its normal boiling point.

BLEVE

28. Explosion in which reaction front moves at a lesser speed than sound in the medium is called ....

Inerting & purging.

29. .... is the calculation of quantitative relationships of the reactants and products in a balanced chemical reaction.

Stoichiometry

30. The temperature of the liquid fuel at which a mixture corresponding to the lean flammability limit is formed due to its vapor mixing with the ambient air is known as ....

Flash point temperature.

31. The distance traveled by a flame before forming a detonation is referred to as ....

Run-up distance.

32. A music concert is being planned to be conducted in an auditorium which has a capacity of 20000 people. The intensity of sound from the loudspeakers is expected to be around 240 dB, which is double the allowable reference intensity. Calculate the sound intensity level.

$$\beta = 10\log\left(\frac{I}{I_0}\right) = 2\,dB$$

33. The incidents results in a release of natural gas. The mass of natural gas within the flammable envelope has been estimated at 1200 kg. The release is into a highly congested region of plant which has overall dimensions of 30 m width, 30 m length, and 8 m height. If the vapor cloud is ignited, at what distance from the edge of the congestion will a peak incident overpressure of 0.004 bar registered?

(Hint: Incident peak overpressure of 0.004 bar occurs at a scaled radius of 25)

The TNT Equivalence Method
From the total mass of fuel (hydrocarbon) in the release (1200 kg) determine

$$\text{TNT equivalence} = \text{TNT equivalence} \times \text{efficiency factor}$$
$$= 10 \times 0.04 = 0.4$$

Equivalent mass of TNT $= 1200 \times 0.4 = 480\text{kg}$

From Figure 3.2 an incident peak overpressure of 0.04 bar occurs at a scaled radius of 25 m, hence the actual radius (r) from the explosion centre is given by:

$$r = 25 \times (480)^{1/3} = 195.7\text{m}$$

34. Calculate the lower flammability limit of a mixture containing 84% methane, 10% ethane, and 6% propane.

$$L_{FL} = \frac{100}{\dfrac{6}{2.1} + \dfrac{10}{3.0} + \dfrac{84}{5.0}} = 4.35\%$$

35. What is meant by stoichiometry? How flammability limits are calculated using stoichiometry?

It is the relationship between the quantities of substances that take part in a reaction or form a compound (typically a ratio of whole integers). It is the calculation of quantitative (measurable) relationships of the reactants and products in a balanced chemical reaction (chemicals).

$$\text{IFL} = \frac{55}{4.76m + 1.19x - 2.38y + 1}$$

$$\text{UFL} = \frac{350}{4.76m + 1.19x - 2.38y + 1}$$

36. Calculate OFSC and ISOC for methane.

Flammability limit in air	LFL: 5.5% fuel in air
	UFL: 18% fuel in air
Flammability limit in pure oxygen	LFL: 5.3% fuel in oxygen
	UFL: 63% fuel in oxygen
LOC	11% oxygen

$$\text{OSFC} = \frac{\text{LFL}}{1 - z[\text{LFL} / 21]} = \frac{\text{LOC}}{z[1 - [\text{LOC} / 21]]} \qquad \text{ISOC} = \frac{zx\,\text{LFL}}{1 - [\text{LFL} / 100]} = \frac{zx\,\text{LOC}}{z - [\text{LOC} / 100]}$$

$$\text{OFSC} = 11.53\% \qquad\qquad\qquad\qquad \text{ISOC} = 11.64\%$$

37. Explain failure rate of a component.

Failure rate follows a typical bath-tub curve. Highest failure rate exhibits for a component at infant mortality stage and old stage; between these two stages, failure rate is reasonably constant. On average, most components fail after certain periods of time. This is called average failure rate (λ) with units faults per time. For constant failure rate (λ) is given by exponential distribution

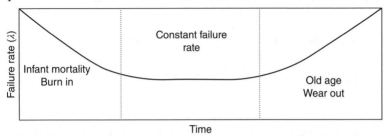

38. What are the different methods used in estimating the explosion damage consequences. Explain any one of them in detail.

Damage is a function of rate of pressure increase and duration of blast wave. Blast damage are estimated based on the peak side-on overpressure.

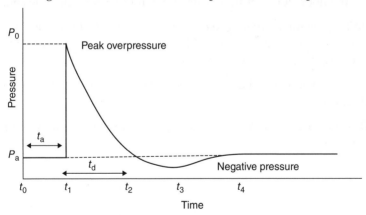

TNT equivalence method

$$m_{TNT} = \frac{\eta m \Delta H_c}{E_{THT}} \qquad Z_e = \frac{r}{\sqrt[3]{m_{TNT}}}$$

39. (a) A pressure vessel of volume $1\,m^3$ is designed to withstand a pressure of $500\,bar$. However, after prolonged use and corrosion at a weld, the pressure vessel fails at a pressure of $450\,bar$ when it is being charged

with compressed air. Determine the TNT equivalence for the explosion based on energy release. (Hint: energy release from TNT is 4520 kJ/kg)

(b) If in the above accidental explosion, an overpressure of 75 kPa was observed at a distance of 5 m from the pressure vessel, find the percentage yield. (Hint: Take mass of TNT as 1.37 kg)

(a) $E = \dfrac{(p - p_o)}{\gamma - 1} V_o = E = \dfrac{(450 - 1)}{1.4 - 1} \times 10^5 \times 1 = 112.25 \, MJ$

Equivalent mass $= \dfrac{112.25 \times 10^3}{4520} = 24.83 \, kg$ of TNT

(b) Percentage yield $= \dfrac{1.37}{24.83} \times 100 = 5.52\%$

40. What are the different types of explosions occurring in oil and gas industries? Explain them briefly.

Confined vapor cloud explosion (CVCE): An explosion in a vessel or a building caused due to release of high pressure or chemical energy.

Vapor cloud explosion (VCE): Explosion caused by instantaneous burning of vapor cloud formed in air due to release of flammable chemicals.

Boiling liquid expanding vapor explosion (BLEVE): Explosion caused due to instantaneous release of large amounts of vapor through narrow opening under pressurized conditions.

Vented explosion (VE): Explosion due to high speed venting of chemicals.

Dust explosion: Explosion resulting from rapid combustion of fine solid particles.

4

Safety Measures in Design and Operation

4.1 Introduction

Safety measures can be adopted in both design and operational stages to avoid catastrophic incidents. One of the major events that can result in serious consequences in oil and gas industries is fire and explosion. With respect to control of accidents in process industry, the major focus is to prevent fire and explosion. While there are many methods to accomplish this, for a fire, accident, or combustion explosion to happen, three basic conditions must be fulfilled: (i) presence of combustive or explosive material; (ii) presence of oxygen to support the combustion reaction; and (iii) source to initiate the reaction. Absence of any one of the three may not induce a blast or burst. It is therefore imperative, though impractical, that one should design a process scheme by avoiding the presence of combustive material. This is due to the basic fact that offshore drilling operations deal with explosive material and a huge inventory of these in their storage compartments. Further, combustion mechanism can never be averted due to the presence of oxygen in abundance; this will rather support or initiate fire. By default, due to the electric processes present in an offshore and petroleum industry, source of kindling is an integral component of the organization. The fundamental question that we need to address is how can we design the system effectively?.

Health, Safety, and Environmental Management in Offshore and Petroleum Engineering, First Edition.
Srinivasan Chandrasekaran.
© 2016 John Wiley & Sons, Ltd. Published 2016 by John Wiley & Sons, Ltd.
Companion website: www.wiley.com/go/chandrasekaran/hse

There are many preventive measures for fire and explosion. One of the common methods is inerting or purging. This is a mechanism by which the oxygen content or the fuel concentration is brought below a target value. For this purpose, flammability diagram is used to control the arms of the fire triangle. Static electricity present in the system can also be controlled using this method. Layout of the plant, with a good ventilation system can also help dilute the flammable mixture with more air concentration. Use of explosion-proof equipments and instruments are strongly recommended as one of the effective hazard control measures for fire and explosion. Well-designed sprinkler systems can also be one of the feasible solutions.

4.2 Inerting or Purging

Inerting or purging is a mechanism by which one can reduce the oxygen or fuel concentration below a specific target value. Usually, this is 4% below the limiting oxygen concentration (LOC). Nitrogen or carbon dioxide can also be used but nitrogen is commonly preferred in the purging process. Purging essentially reduces the oxygen concentration in the environment. Various methods are available for purging such as vacuum purging, pressure purging, combined purging, vacuum and pressure purging with impure nitrogen, sweep-through purging, and siphon purging.

4.3 Terminologies

Flammable Limits
> They are the lowest (lower limit) and highest (upper limit) concentrations of a specific gas or vapor in mixture with air that can be ignited at ordinary temperature and pressure.

Holding Purge
> The procedure of maintaining an inert gas or liquid in a closed system during maintenance or repair, which has been introduced to replace the normal combustible content.

Hot Cutting
> Cutting by oxyacetylene torch or by other means into any pipeline or a vessel containing only combustible gas at slightly above atmospheric pressure.

Hot Tap
> Cutting into a pipeline containing a combustible gas or liquid by use of a special machine. The machine is attached to suitable fittings,

which have been previously welded or otherwise secured to the loaded pipeline. The tapping machine and fittings are constructed such that the required size of the opening may be cut in the loaded pipeline and the machine may be safely removed without appreciable loss of combustibles.

Inert Gas

A gas, noncombustible and incapable of supporting combustion, which contains lesser than 2% oxygen concentration and combustible constituents lesser than 50% of the lower explosive limit of the combustible being purged.

Inert Purge

The act of changing the contents of a pipe or container by using an inert substance to displace the original content or to separate the two media from being interchanged. Flammable mixtures are thus avoided.

Purge

The act of removing the contents of a pipe or container and replacing it with another gas or liquid.

Purge Gas

Gas used to displace the contents of a container. To avoid flammable mixtures, purge gas is usually inert except in certain instances where a relatively smaller area of contact allows the amount of flammable mixture to be minimized and controlled satisfactorily.

Purge into Service

The act of replacing air or inert gas in a closed system by a combustible gas, vapor, or liquid.

Purge Out of Service

The act of replacing the normal combustible content of a closed system by inert gas, air, or water.

Pyrophoric

A substance or mixture that can ignite spontaneously.

Slug

A quantity of inert gas interposed between combustible gas and air during purging. Slug does not fill the complete length of the pipe but moves through the pipe as a separate mass to prevent mixing of gas and air.

Stratification

It is a process by which different gases settle into layers.

Ventilation

The process by which doors, manholes, valves, etc. are opened to permit the ingress of air by natural circulation. This helps in replacing the gas contents.

4.4 Factors Affecting Purging

Replacement of one gas by another in an enclosed space or chamber takes place by means of two distinct actions: (i) displacement; and (ii) dilution or mixing. In a purge that is effected entirely by displacement; gas or air that is originally present in the container is pushed out of the escape vents by the entry of purge gas with little or no mixing of the purge gas with original contents. Thus, quantity of purge gas required for purging by displacement approximates the quantity of gas or air replaced. Certain conditions such as the size or shape of the chamber and the nature of gases cause the purge gas to mix with the original contents; this may lead to the dilution of purge gas. Purging by dilution can be accomplished in some situations by alternately pressurizing and depressurizing the facility. To accomplish a satisfactory purge by dilution or mixing, volume of inert purge gas that is required may be four or five times that of the free space of the chamber being purged. Increased demand in its contents is due to the fact that as purging proceeds, increasing amount of purge gas is lost from the escape vents in mixture with the original contents.

Almost all purging operations are combinations of displacement and dilution actions. In the actual practice, it is impossible to avoid mixing of the purge gas with air that is being replaced. In general, lesser the mixing or dilution more is the efficiency of purge. Purging, which proceeds with mixing or dilution that occurs in tanks and holders should be accomplished with an inert purge medium to avoid flammable mixtures. Purging without the use of an inert medium should be limited to pipelines where the amount of mixing can be controlled satisfactory by other methods.

4.5 Causes of Dilution or Mixing

Factors affecting the relative proportions of displacement and mixing action in a purge should be understood thoroughly so that careful attention can be given to avoid or minimize those factors or conditions that promote mixing. Some of the more important causes of mixing during a purging are: (i) large area of contact that promotes natural diffusion; (ii) long periods of contact that permits natural diffusion; (iii) agitation resulting from a high input velocity; (iv) gravitational effects resulting from the introduction of heavy gas over a light gas or vice versa; and (v) temperature changes and differentials causing convection currents. Failure to recognize the importance of parameters such as location of the purge gas input connection, rate of input of the purge gas, and the temperature differentials can result in a purging operation, which is 80–85% dilution and only 15–20% displacement.

4.5.1 Area of Contact

There is always some diffusion of the purge gas into the original gas and of the latter into the purge gas at the surface of contact. Amount of mixing those results from diffusion depends on the area of contact. This in turn depends on size, shape, and internal construction of the chamber being purged. Contact area has a very significant effect on the efficiency of a purge. When purging a tall, narrow tower, the area of contact between the gases is smaller in comparison to their volumes. Mixing is therefore limited and the quantity of inert purge gas used may not be greater than the volume of gas or air to be cleared out.

The crown of a storage holder, in contrast, is a flat, shallow dome, the height of which is significantly less than the diameter. It is impossible to avoid having a very large area of contact in a chamber of this shape. Consequently, it is usually necessary to use at least 1.5–2.5 volumes of inert gas per volume of free space in purging. When purging a pipeline, the area of contact may be so small that little mixing will occur. One can take advantage of this condition to conduct an inert purge by using a quantity of inert gas that is only a fraction of the volume of combustible gas or air to be replaced. It is possible to introduce just enough inert gas to form a "slug" or piston between the original gas (or air) content and the air entering (or gas). This slug and the original gas or air ahead of it is pushed along the pipe to the end of the section being purged by air or gas introduced after it. Recent research has greatly expanded the understanding of the slugging process, particularly for larger diameter pipelines.

4.5.2 Time of Contact

The duration of contact of the surfaces of purge gas and original gas or air should be as short as possible. If the purge gas input rate is too low it will result in excessive mixing by natural diffusion. Interruptions and variations of the purge gas input rate should be avoided as far as possible.

4.5.3 Input Velocities

Velocity of the purge gas at the entrance plays a significant effect on the nature of purge. In general, size of the inlet of the purge gas to containers other than pipe should be as large as practical so that the input velocity does not exceed

0.6–1 m/s. This keeps the agitation or stirring of the chamber contents at a minimum. If the input connection is relatively small in comparison to the rate of input, it may result in higher velocity. This will carry the purge gas up to the center or across the chamber, resulting in thorough mixing. If the input velocity is higher and the outlet vent is larger, purge gas may stream or arc across from the inlet to the outlet, limiting both displacement and dilution.

4.5.4 Densities of Gases

Relative densities of the purge gas and the gas (or air) being purged have important effects on the mechanics of the purging action. For example, carbon dioxide has a specific gravity of approximately 1.5. This is large enough when compared to that of a natural gas, which is approximately 0.6. This will create the inert gases to stratify and remain in a layer at the bottom of the chamber filled with natural gas. Therefore, when purging a light gas out of a chamber, an effort is made to push the lighter gas out through vents in the top of the chamber by allowing the heavier gas at the base. Conversely, in putting equipment back into service, when heavier inert gas is to be replaced by lighter ones, the latter should be introduced at the top of the chamber, while the heavier gas is vented from the bottom.

When purging facilities out of service that have contained gases with a higher specific gravity, vapors can be effectively replaced with a minimum of mixing by introducing the inert gas at the top of the chamber and displacing the vapors downward through bottom vents. When purging facilities into service that are to receive such substances and after replacement of the air by an inert purge gas, heavy vapors or liquid should be admitted at the base of the vessel while the purge gas is displaced upward through the top vents. Heavier gases such as butane, propane, or benzyl vapors can first be displaced downward. Natural gas is displaced upward and out of the top vents by an inert gas, which is subsequently replaced by air. The importance of differences in densities in a purging operation demands that about 50% of inert gas is to be replaced by air in a large chamber. A purge gas such as nitrogen has a specific gravity of approximately 0.97, which is almost identical to that of air so that mixing is not restrained by stratification when natural gas is being replaced.

4.5.5 Temperature Effects

It is desirable to keep the temperature of the purge gas entering a large chamber as low as practicable to minimize the possibility of setting up any "thermal currents." A positive pressure must be maintained within a chamber being purged. Thus, when a sudden drop in atmospheric temperature occurs during purging of a vessel, it may be necessary to reduce the rate of release of the purged gas (or air). This is required to offset the contraction of the contents present in the chamber. However, it is not necessary to control temperatures when the chamber being purged contains deposited solids or liquids. Special precautions should be taken if the holder, tank, pipe, or other facility contains naphthalene or tar deposits, oils, solvents, or other materials that will volatilize and give off combustible vapors even under marginal increase in temperature above the ambient value. Either before or during the purge, these deposits should be heated to such a degree that there could be no further volatilization of combustible vapors when air is admitted to the chamber. This topping distillation of deposits may be accomplished by a few operations: (i) steaming of the chamber or system prior to gas purging; (ii) using steam as the purging gas; or (iii) admitting the purge gas at an elevated temperature, saturated with water vapor. To attain a high mechanical efficiency during purging, it is necessary to keep mixing and dilution at a minimum by: (i) avoiding interruptions or variations in purge gas input; (ii) using large input connections; (iii) controlling the input velocity; (iv) introducing purge gas at a proper location with respect to gas densities; (v) avoiding differences and sharp changes in temperature; or (vi) using larger vents so that ready escape of displaced gas is possible.

4.6 Methods of Purging

There are many methods of purging that are commonly practiced in process industries. Broadly there are two types namely: batch purging and continuous purging. Some of the common methods of purging are as follows:

4.6.1 Siphon Purging

In this method, equipment that is to be purged is filled with liquid. Purged gas is introduced into the vapor space to replace the liquid, which is drained from the enclosure. Volume of the purge gas required in this method will be equal to the volume of the vessel and the rate of application will be the same as the rate of draining.

4.6.2 Vacuum Purging

In this method, equipment that normally operates at reduced pressure (or capable of operating at reduced pressure) is purged during the shutdown by breaking vacuum with the purge gas. If the initial pressure is not lower enough to ensure the desired low oxidant concentration, it becomes necessary to reevacuate and repeat the process. Amount of the purge gas required in this case is determined by the number of applications required to develop the desired oxidant concentration. Where two or more containers or tanks are joined by a manifold and to be purged as a group, then the vapor content of each container or tank should be checked for completeness of purging.

4.6.3 Pressure Purging

In this method, enclosures might be purged by increasing the pressure within the enclosure. This is carried out by introducing the purge gas under pressure. After the gas has diffused, the enclosure is vented out to the atmosphere. In this case, more than one pressure cycle might be necessary to reduce the oxidant content to the desired percentage. Where two or more containers or tanks are joined by a manifold and to be purged as a group, then the vapor content of each container or tank should be checked to determine that the desired purging has been accomplished. Where a container filled with combustible material is to be emptied and then purged, purge gas is applied to the vapor space at a pressure consistent with equipment design limitations, thus accomplishing both the emptying of the vessel and the purging of the vapor space in the same process in parallel.

4.6.4 Sweep-Through Purging

This method involves introducing a purge gas into the equipment at one opening and letting the enclosure content escape to the atmosphere through another opening, thus sweeping out residual vapor. The quantity of purge gas required depends on the physical arrangement. Pipelines can be effectively purged with a little more volume of purge gas if the gas can be introduced at one end and the mixture can be released at the other in parallel. However, vessels require quantities of purge gas much in excess of their volume. If the system is complex, involving side branches through which circulation cannot be established, the sweep-through purging method might be impractical; in such cases pressure or vacuum purging is more appropriate. In this method, the total quantity required might be lesser than that for a series of steps of pressure purging. Further, four to five times of volumes of purge gas is required to almost completely displace the original mixture for ensuring a complete mixture.

4.6.5 Fixed-Rate Purging

This method involves a continuous introduction of purge gas into the enclosure at a constant rate. This should be sufficient to supply the peak requirements so that that complete protection is provided. The advantages of this method are simplicity, lack of dependence on devices such as pressure regulators, and possible reduced maintenance. The main disadvantage is the continuous loss of product where the space contains a volatile liquid due to constant "sweeping" of the vapor space by the purge gas. Other demerits are increased total quantity of purge gas since it is supplied regardless of whether it is needed or not. Possible disposal problems that arise from the toxic and other effects are also few other undesirable consequences of this method.

4.6.6 Variable-Rate or Demand Purging

This method involves the introduction of purge gas into an enclosure at a variable rate that is dependent on demand. The advantage is that the purge gas is supplied only when actually needed. Therefore it is possible to completely prevent influx of air, if desired. The disadvantage is that the operation depends on the functioning of pressure control valves; however, it is difficult to maintain these valves as they sometimes operate at low pressure differentials.

4.7 Limits of Flammability of Gas Mixtures

One of the basic requirements in a purging operation is the knowledge of the flammable limits of combustible gas in air. When small increments of a combustible gas are progressively mixed with air, a concentration is finally attained in which a flame will propagate in the presence of source of ignition. This is referred to as the lower flammable limit of the gas in air. For practical purposes, this is considered the same as the Lower Explosive Limit. As further increments of gas are added, a higher concentration of flammable gas in air will finally be attained in which a flame will fail to propagate. The concentration of gas and air at this point is referred to as the upper flammable limit of the gas in air. For practical purposes, the upper flammable limit is considered the same as that of the Upper Explosive Limit (UEL). Flammable gas–oxidant mixture might be controlled by reducing the concentration of oxidant or by adding an inert constituent to the mixture. Both processes can be explained most easily by referring to a flammability diagram. An increase in temperature has a similar effect on the flammability diagram. The exact effects on a system produced by changes in pressure or temperature should be determined for each system.

4.8 Protection System Design and Operation

The owner or operator shall be responsible for a thorough analysis of the process to determine the type and degree of deflagration hazards inherently present in the process. Information required for the oxidant concentration monitoring and control shall be compiled and documented carefully. This includes factors, such as: (i) monitoring and control objectives; (ii) monitored and controlled areas of the process; (iii) dimensioned drawings of the process. To design the protection system, additional information required with respect to the existing design includes: (i) equipment make and model, if available, including its design strengths; (ii) plan and elevation views with flows indicated; (iii) startup, normal, shutdown, temporary operations, and emergency shutdown process conditions and ranges for the following factors:

(a) Flow
(b) Temperature
(c) Pressure
(d) Oxidant concentration
(e) Process flow diagram and description
(f) Ambient temperature in process area
(g) Process interlocks

It is obligatory on the part of the owner or operator to disclose all process information required for the protection system design and also to educate the employees through periodic training programs.

4.9 Explosion Prevention Systems

The owner or operator shall be responsible for the maintenance of the system after installation and acceptance based on procedures provided by the vendor. They are also responsible for the periodic inspection of the system by the authorized personnel. All documentation relevant to the protection system should be retained in accordance.

4.10 Safe Work Practices

A few important work practices, which could result in a safe and healthy work environment, are listed as follows:

4.10.1 Load Lifting

The manufacturer's rated load capacity shall not be exceeded on cranes or other load-lifting devices. These equipments should be operated and maintained in accordance with manufacturer's recommendations. Tag lines should be used to guide and steady all loads being lifted.

4.10.2 Confined Space, Excavations, and Hazardous Environments

Where hydrogen sulfide, sulfur dioxide, carbon dioxide, or other hazardous gas is present or suspected to exist, the operator shall ensure that all personnel, contractor, and service company supervisors are advised of the potential hazards. Safety guidelines and recommendations for use in the production operations where hydrogen sulfide or sulfur dioxide gas may be encountered should be followed as per the recommended guidelines and practices (API RP 55; ANSI 2117.1). A confined space is an area, which either has an adequate size and configuration for people to enter or has limited means of entry or exit or is not designed for continuous employee occupancy should be carefully examined and notified. Examples of confined spaces that are commonly found at onshore producing facilities are well cellars, electrical vaults; fin fan coolers, tanks, vessels, diked areas, and valve pits. Confined space hazards should be identified for all facilities in the workplace and safe work practices should be established for working in such confined spaces. A confined space entry permit shall be used to enter any confined space that has atmospheric, engulfment, or configuration hazards. Attendant and emergency rescue services shall be provided for all permit-required confined spaces. When preparing the confined space for entry, precautions must be in place to ensure that the space remains safe for operation/maintenance. This may include forced air ventilation, equipment isolation, or other measures that are deemed necessary for ensuring operational safety. For equipment isolation, blinding, double block and bleed, or other equipment and energy isolation controls should be considered.

4.10.3 Lockout/Tagout

A lockout/tagout program shall be established to control hazardous energy. They consist of the following measures:

- Locks and/or tags should be placed to clearly identify the equipment or circuits being worked on.
- Systems locks or tags should include the identity or job title of the person who installed the lock/tag.
- Personnel should be trained and disciplined in the use of this system to prevent unexpected operation of any equipment that stores any type of energy that might cause injury to the personnel.
- The lock or tag should be removed by the person who installed it. In the event the individual is not available, the lock or tag may be removed by

the supervisor after ensuring that no hazard will be created by energizing the locked or tagged equipment or circuit(s).

4.10.4 Well Pumping Units

Power to the pumping units should be deenergized and locked or tagged out to eliminate potential hazards during well servicing operations. In the concerned locations, overhead electric power to the pumping unit control panel should be deenergized. Wherever necessary, power service should be deenergized while moving the rig in or out and during rig-up and rig-down operations. During well servicing operations, pumping units should be secured to prevent unintended movement. Chain or the sling of the wire rope of suitable strength should be used to handle the horse head if removal or installation operations are necessary. On installation, the horse head should be bolted or latched in accordance with the manufacturer's specifications. Upon completion of the well servicing operations and before reenergizing the power source, precautions should be taken to ensure that all personnel and equipments are clear off the weight and the beam movement. Brake systems on all pumping units in service should be maintained in safe working condition. After well servicing operations are completed, all pumping unit guards and enclosure guards (belt and motor sheaves) should be reinstalled prior to the startup of the well. Guards need not be in place until all panel adjustments (pump, spacing, etc.) are examined for safe operations.

4.11 Hot Work Permit

A written safe work permit (hot work permit) system, covering welding and flame cutting operations should be observed in case of working at special jobs. In general, a safe work permit system should consist of authorization to do the work along with the following:

(a) Pre-Work stage communication meetings addressing the following:
   - Simultaneous operations
   - Air/gas testing
   - Equipment isolation
   - Equipment preparation
   - Identification of hazards
   - Emergency procedures

(b) Work-in-progress stage:
  • Posting of permit
  • Air/gas testing
  • Personal protective equipment requirements
  • Fire watch
  • Special procedures/precautions
(c) Return to Service Stage
  The supervisor should hold a pre-job meeting with the crew and other involved persons to review responsibilities for the operation to be performed. Welding and flame-cutting operations should not be permitted close to flammable liquids, accumulations of crude oil, escaping gas, or locations where sparks, flames, heat, or hot slag could be sources of ignition. Only qualified persons should perform welding or flame-cutting operations on equipments used to contain hydrocarbons or hazardous materials. Appropriate personal protective equipments should be used for hot work operations. If the object to be cut or welded cannot be readily moved, all movable pre-hazards in the near vicinity should be taken to a safe place. If they cannot be taken to a safe place, guards should be used to confine the heat, sparks, and slag to protect the immovable pre-hazards. In such cases, a safe welding area may be designated. In this area, welding and flame-cutting operations can be conducted with minimal concern of providing an ignition source for flammable hydrocarbons or combustible materials.

A safety work permit is not normally required for welding operations in an approved safe welding area, provided properly maintained pre-extinguishing equipments are available for immediate use. A minimum of at least one 30-lb dry chemical pre-extinguisher should be immediately available during welding or cutting operations in addition to the general pre-protection equipment. Fire watches with extinguishing equipments should be placed whenever welding or cutting is performed in locations other than designated safe welding areas. A pre-watch should be maintained for at least one hour after the completion of the welding or cutting operations in such areas. Before cutting or welding is permitted in areas outside a designated safe welding area, the area should be inspected by the individual who is responsible for authorizing cutting or welding operations. This individual should designate the precautions to be followed in granting authorization to proceed. Cutting or welding should not be permitted in the following situations: (i) areas that are not authorized by the supervisor; (ii) in the presence of an explosive atmosphere; (iii) areas near the storage of large quantities of exposed readily ignitable materials; (iv) areas where ignition can be caused by heat conduction, such as metal walls or pipes that are in contact with combustibles on the other side; and (v) on used containers such as drums unless properly cleaned.

4.12 Welding Fumes and Ventilation

During welding operations, toxicity depends on the composition and concentration of the fumes. Composition and quantity of fumes depends on the materials being welded, composition of the welding rods, any coatings or paints encountered in the welding operations, the process used, and the circumstances of use. Toxic fumes can be generated from welding on metals that are coated with (or) contain alloys of lead, zinc, cadmium, beryllium, and other similar metals. Some paints and cleaning compounds may also produce toxic fumes when heated. Potential health effects that arise from welding depend on the type and severity. Where the eyes or body of personnel may be exposed to injurious materials, eyewash and shower equipments for emergency use should be provided.

4.13 Critical Equipments

A critical equipment is defined as the one that is essential in preventing the occurrence of (or mitigating) the consequences of an uncontrolled event. Such equipments include pressure vessels, pressure relief devices, compressors, alarms, interlocks, and emergency shutdown systems. Critical equipments should be periodically inspected and tested as recommended by the manufacturer or in accordance with the recognized engineering practices. When using Nondestructive Testing (NDT) methods, qualified personnel should conduct the tests in accordance with the recognized methodology and acceptance criteria. Most importantly, when the critical equipment is removed from service, a program should be in place to ensure that an equivalent protection is provided.

4.13.1 Changes to Critical Equipment

Procedures to manage changes to critical equipment should be implemented as appropriate. These procedures should address the following prior to making the change:

1. Basis for the proposed change
2. Impact of change on safety and health

3. Modifications to operating procedures
4. Authorization requirements for the proposed change

Employees whose job tasks will be affected by the change in the critical equipment should be informed of the change prior to start-up.

4.14 Fire Prevention

Safe storage of combustible and flammable materials and their appropriate location are important to ensure fire protection. Smoking is prohibited in the vicinity of the operations that constitute a pre-hazard. Such locations should be conspicuously posted with a sign board of *no smoking*. Changing rooms and other buildings where smoking is permitted should be located in areas designated safe for smoking. No source of ignition should be permitted in an area where smoking has been prohibited, unless it is first determined safe to do so by the supervisor in-charge or his designated representative. Potential sources of ignition should be permitted only in the designated areas located at a safe distance from the Well head or flammable liquid storage areas. Equipment, cellars, ground areas around and adjacent to the facility should be kept free from oil and gas accumulations that might create or aggravate pre-hazards. Combustible materials such as oily rags and waste should be stored in the covered metal containers. Natural gas or liquefied petroleum gas should not be used to operate spray guns or pneumatic tools.

4.15 Fire Protection

Fire-fighting equipments should not be removed for other than pre-protection, pre-fighting purposes and services. A pre-fighting water system should be used for wash down and other utility purposes so long as its pre-fighting capability is not compromised. Fire extinguishers and other pre-fighting equipments should be suitably located, readily accessible, and clearly labeled by their type and method of operation. Fire suppression equipments such as extinguishers, fixed systems etc. should be periodically inspected and maintained in operating condition at all times. A record of the most recent equipment inspection should be maintained. Portable pre-extinguishers shall be tagged with a durable tag showing the date of the last inspection, maintenance, recharge using the acceptable record keeping media. Inspection and maintenance procedures should comply with NFPA 10. Personnel should be familiar with the location of pre-control and selected personnel trained in the use of such equipments. Fire-fighting equipment shall be accessible and free of obstructions.

4.16 Grounding and Bonding

Production facilities are subject to various forms of electrical hazards that must be protected against. Static electricity can be generated by fluid movement inside vessels, pipes, and tanks. This results in generation of static sparks, which can be a potential ignition source; lightning striking a facility is also another potential ignition source. It is important to note that the failure of electrical equipments can occur when exposed to shock hazards. All equipments should be inspected and maintained according to the guidelines of API RP 2003 and NFPA 77.

4.17 Other General Requirements

Deflagration prevention and control for occupied enclosures should prevent the structural failure of the enclosure and minimize injury to the personnel in adjacent areas outside of the enclosure. Deflagration prevention and control for unoccupied enclosures should prevent the rupture of the enclosure. They should be arranged to avoid injury to the personnel and designed to limit the damage of the protected enclosure. They should be designed to avoid damage to the adjacent properties in the near vicinity. The design basis of deflagration hazard scenario should be identified and documented to obtain an acceptable risk level as permitted by the competent authority of the local jurisdiction. Compliance options are of two types: (i) performance-based design; and (ii) prescriptive-based design.

4.17.1 Performance-Based Design

To continue meeting the performance goals and objectives of the standards set by the oil and gas industries, the design features, required for each prevention and control system, should be maintained for the life of the protected enclosure. Any change to the design requires an approval of the authority having jurisdiction prior to the actual change. The design of prevention and control system should be based on the documented hazard scenario. It should be ensured that the combustible material outside the enclosure does not attain their ignition temperature from flame or hot gases. Prevention and control systems should limit the risk of damage to exposed structures and does not expose personnel to flame, hot gases, hot particles, toxic materials, or projectiles. They should be designed such that they limit the risk of flame spread from vessel to vessel via interconnected ducts and thus avoid cascading damages. Inspection and maintenance of these equipments should be documented and retained for at least one year or last three inspections. The owner or operator should be responsible for a thorough analysis of the process to determine the type and degree of deflagration hazards inherent in the process. Information required for the oxidant concentration monitoring and control should be compiled and documented.

The owner or operator shall disclose any and all process information required for the protection system design. A design record file including data sheets, installation details, and design calculations should be assembled following the requirements and maintained for each application, suitable for use in validating the system design including the following criteria:

- Manufacturer's data sheets and instruction manuals
- Design calculations including final reduced pressures
- General specifications
- Explosion prevention system equipment list
- Sequence of operation for each system
- End-user inspection and maintenance forms
- User documentation of conformity with applicable standards and the appropriate chapter of this standard
- Combustible material properties and source of data
- Process hazard review
- Process plan view including protected process, placement location of all explosion prevention devices, and personnel work locations
- Process elevation view
- Electrical wiring diagram, including process interlock connection details
- Mechanical installation drawings and details
- Electrical installation drawings and details
- Process interlocks identifying each equipment interlock and function (P&ID)
- Employee training requirements

All the design and installation parameters shall be field-verified prior to their installation. As-built drawings, user instruction manuals, and service maintenance requirements should be presented to the owner or operator at the completion of the project. Mounting locations should follow the manufacturer's requirements, since explosion prevention systems are location sensitive. Location changes should be made only with the approval of the explosion prevention system manufacturer. Mounting locations are chosen so as not to exceed maximum operating temperatures of system components. Mounting locations include safe access for installation, service, inspection, and maintenance, up to and including work platforms as required by local workplace safety regulations. Detectors shall be mounted according to manufacturer instructions to protect them from shock, vibration, accumulation of foreign material, and clogging or obscuration of the sensing area. Discharge nozzles should be located and oriented so that they will not be obstructed by the structural elements in the discharge pattern or by the solid particle accumulation.

Mechanical installation and system components should be fabricated using materials that are free from corrosion and other contaminants. Detectors should be mounted in such a way that it inspects and removes obstructions to the sensing pathway. Detector mounting should incorporate means to minimize the unwanted system actuation due to vibration or shock, wherever applicable. Mounting hardware and the mounting surfaces for all protection system components should be capable of withstanding the static and dynamic loads, including the thrust or impulse pressure that arises from the temperature requirements during process. Agent storage containers that are installed externally to the protected process shall be mounted in such a manner that they can be easily inspected and free from obstructions in the pathway. All electrical equipment and installations shall comply with the requirements of NFPA 70. Terminals and connections should be protected from moisture and other contaminants. Hazardous areas that are identified should be documented and maintained on file for the life of the facility. Wiring for input and output control circuits should be isolated, shielded, and protected from other wiring to prevent possible induced currents. Instrumentation included as part of the explosion prevention or protection systems should meet the recommendations of the oil industry. Control systems that are meant for emergency activation should be installed, maintained, and isolated from the basic process control systems. When supported by a documented hazard analysis, the functional testing of all the requirements shall be carried out as part of routine inspections. National Fire Alarm Code shall be employed when measurement devices, explosion detection devices, controlling valves, releasing devices, solenoids, actuators, other supervisory devices that monitor critical elements or functions such as low pressure switches, notification appliances, and signaling line circuits are connected to the control panel. It is important to note that a signaling line circuit that is used as a part of an explosion protection or suppression system shall not be shared by any other operating systems. It should not be used by more than one explosion prevention system unless certified by the original manufacturer. Prior to use, factory authorized personnel shall check the explosion prevention system, which includes: (i) conducting a walkthrough and general visual inspection of correct location, size, type, and mounting of all system components; (ii) physical inspection of the system components, including mechanical and electrical component integrity; (iii) conducting the functional testing of the control unit; (iv) point-to-point wiring checks of all circuits; (v) ensuring continuity and condition of all field wiring; and (vi) inspecting the sensing pathway and calibrating the initiating devices. In addition to the above, following aspects of testing are also implemented:

- To verify the correct installation of system components including sensing devices, fast-acting valves, suppressant storage containers, nozzles, spreader hoses, protective blow-off caps, plugs, and stoppers.
- To verify the system sequence of operations by simulated activation to verify system inputs and outputs.
- To conduct automatic fast-acting valve stroke test.
- To conduct prevalidation testing, verify system interlocks, and shutdown circuits.
- To identify and fix discrepancies before arming and handing off to a user or operator.
- To recalibrate detection sensing devices to final set points.
- To complete record of system commissioning inspection including hardware serial numbers, detector pressure.
- To calibrate the suppressor and valve actuators charging pressures (psig), as appropriate.
- To conduct end-user training.
- To conduct final validation testing of all equipments under the guidance of competent authority.
- To arm the explosion prevention system.

4.17.2 Inspection of Protection Systems

Protection and control systems should be inspected and tested at a periodic interval of 3 months. While systems designed by the owner or operator shall be inspected by the authorized personnel, those designed by the manufacturer shall be inspected by the trained personnel. Maximum inspection and test interval should not exceed 2 years. An inspection of explosion prevention systems shall be conducted in accordance with the system designer's requirements and project specifications. This shall include the following conditions, wherever applicable:

- The process and processed material have not changed since the last inspection.
- The explosion prevention system has been properly installed in accordance with this standard and the manufacturer's instructions.
- System components, including mounting arrangements, are not corroded or mechanically damaged.
- User operation instructions are provided near the control unit.
- System components are clearly identified as an explosion prevention device.
- System components have no damage from the process, acts of nature, or debris.
- System components have not been painted or coated without prior approval from the original equipment manufacturer.
- System components are not blocked by process material.
- System components have not been tampered with.
- The system has not discharged or released.
- System seals, tamper indicators, or discharge indicators, if provided, are in place and functioning.
- The control unit functions according to design requirements, circuits are properly supervising the system and status is "normal condition" when armed.
- The system wiring is free from ground conditions and faults.
- System suppressors and valve actuators are pressurized and operational.
- System interlocks are verified for proper sequence and functioning.
- Mechanical isolation, if used (such as rotary valves, etc.) is maintained within the requirements of this standard and design tolerances.
- Plant fire notification is verified.
- System sequence of operation is verified by simulated activation.
- System component serial numbers are verified as the same as those recorded during the last inspection.

4.18 Process Safety Management (PSM) at Oil and Gas Operations

Generally, PSM standards emphasize the implementation of a systematic and potentially complex program to identify, evaluate, prevent, and respond to releases of hazardous chemicals in the workplace.

4.18.1 Exemptions of PSM Standards in Oil and Gas Industries

PSM standards include some key exceptions for oil and gas operations. As a major setback, PSM standards exempt oil and gas well servicing because Occupational Safety and Health Administration(OSHA), U.S. Dept. of Labour had begun a separate rulemaking due to the unique nature of those operations. OSHA dropped that rulemaking from its regulatory agenda before the agency promulgated a final rule, but the PSM exemption remained. Following the concerns raised by the American Petroleum Institute about the absence of an economic analysis of the PSM standard as applied to oil and gas production facilities (which, according to OSHA, begins at the top of the well), OSHA suspended enforcement of the PSM standards for oil and gas production operations. PSM standards also exclude flammable liquids stored in atmospheric storage tanks. PSM standards also exempt the normally unoccupied remote facilities. The above exemptions, which are justifiable under practical conditions had limited the applicability of PSM standards to oil and gas exploration and production operations.

More broadly, many industry groups state that oil and gas facilities are already in strict compliance with numerous safety standards tailored to those operations even in the absence of a specific PSM requirements. For example, while most atmospheric storage tanks in use at oil and gas facilities are currently exempted from the PSM standards, they must comply with OSHA's flammable liquids standards as well as other state and local safety standards. Thus, subjecting oil and gas facilities to the PSM standards would mostly serve to add layers of compliance complexity without a commensurate safety benefit. The US Environmental Protection Agency (EPA) has recently placed a greater emphasis on examining the oil and gas facilities to their compliance with the "general duty clause" of the Clean Air Act's risk management program. EPA enforcement has led to several consent orders in instances where the agency felt that the facility in question had not appropriately identified the accidental release hazards or designed and maintained a safe facility. This renewed focus on the release prevention requirements at oil and gas facilities has resulted in a significant improvement on process safety in oil and gas industries.

4.18.2 Process Safety Information

According to the schedule in the initial assessment-process hazard analysis, every employer must complete a compilation of written Process Safety Information (PSI) before conducting any process hazard analysis. This compilation is to enable the employer and the employees involved in process operations to identify and understand the hazards posed by those processes involving highly hazardous chemicals. The PSI must include the following:

* Information pertaining to the hazards of the highly hazardous chemicals used or produced by the process;
* Information pertaining to the technology of the process.
* Information pertaining to the equipment in the process.

4.19 Process Hazard Analysis (PHA)

Every employer must perform an initial PHA on the covered processes. PHA studies should account for various complexities of the process and must identify, evaluate, and control the hazards involved in the process. The employer must use one or more of the following methodologies:

* What-if
* Checklist
* What-if/checklist
* Hazard and Operability Study (HAZOP)
* Failure mode and effects analysis
* Fault tree analysis
* An appropriate equivalent methodology

The PHA must address hazards present in the process while identifying any previous incident, which had a likely potential for catastrophic consequences in the workplace. Engineering and administrative controls are applicable to the hazards and their interrelationships such as appropriate application of detection methodologies to provide early warning of releases. A qualitative assessment of facility sitting, human factors, and a range of the consequences of the failure of engineering and administrative controls need to be an inherent part of PHA.

Process hazard analysis must be performed by a team with expertise in engineering and process operations. The team must include at least one employee who has experience and knowledge of the process being evaluated. In addition, one of the team members must be knowledgeable in the specific PHA methodology being used. The employer must establish a system to promptly address the team's findings and recommendations; assure that the recommendations are resolved in a timely manner and that the resolution is documented. Once a PHA is carried out and the report is prepared, it becomes a legal binding on the employer to implement the recommendations, at least in a phased manner. It is also mandatory for the employer to document the action taken on each recommendation and submit to safety audit every year for approval. At least, once in every 5 years, PHA must be updated and revalidated by a team to assure that it is consistent with the current process. Employers must retain process hazard analyses and updates or revalidations for each covered process for the life of the process.

4.20 Safe Operating Procedures

The employer must develop and implement written operating procedures that provide clear instructions for safely conducting activities involved in each covered process. This should be consistent with the PSI and must address the following elements:

- Steps for each operating phase, including initial startup; normal, temporary and emergency operations; normal and emergency shutdown; and startup following a turnaround or after an emergency shutdown;
- Operating limits, including consequences of deviation and steps required to correct or avoid deviation;
- Safety and health considerations, including hazards posed by the chemicals used in the process;
- Precautions that are necessary to prevent exposure;
- Control measures that are to be taken if physical contact or airborne exposure occurs;
- Quality control for raw materials and control of hazardous chemical inventory levels;
- Mention of any special or unique hazards that may arise during the process;
- Safety systems and their functions.

Operating procedures must be readily accessible to employees and must be reviewed as often as necessary to assure that they reflect current operating practices. This should include the changes that result from the deviations in the process layout, chemicals, technology and equipment, and changes made to the facilities in due course of time. The employer must certify annually that these operating procedures are current and accurate.

4.21 Safe Work Practice Procedures

The employer must develop and implement safe work practices to provide sufficient control of hazards during operations such as lockout/tagout, confined space entry, opening process equipment, or piping and control over entrance into a facility by the maintenance contractor or other support personnel. These safe work practices apply to all the employees of the organization and contractor employees without any exemption.

4.21.1 Training

Each employee who is presently involved in the process operations or those who are likely to be exposed to these process operations must be trained in an overview of the process and in the operating procedures. Training must lay emphasis on the specific safety and health hazards, emergency operations including shutdown and safe work practices as applicable to the employee's job tasks. Refresher training must be provided at least once in every 3 years, and more often if necessary. Outcome of such trainings should be measured by evaluating the employees through rigorous tests and practical assessment to ensure that he understands and adheres to the current operating procedures of the process. The employer must prepare a record, which contains the identity of the employee, date of training, and the means used to verify if the employee has understood the training.

4.21.2 Pre-startup Review

The employer must perform a pre-startup safety review for new facilities and modified facilities when the modification is significant enough to cause substantial changes in the PSI. The pre-startup safety review must confirm the following prior to the introduction of highly hazardous chemicals to a process:

- Construction and equipment are in accordance with design specifications;
- Adequate safety, operating, maintenance, and emergency procedures are in place;
- For new facilities, a PHA has been performed and recommendations have been resolved or implemented before startup;
- Modified facilities meet the requirements of management of change procedures;
- Training of each employee involved in operating a process has been completed.

4.22 Mechanical Integrity

The employer must establish and implement written procedures to maintain the ongoing mechanical integrity of process equipments. The employer must train each employee involved in maintaining the ongoing integrity of process equipments in an overview of that process and its hazards. Training should also address the procedures applicable to the employee's job tasks to assure that the employee can perform the job tasks in a safe manner.

This applies to the following process equipments:

- Pressure vessels and storage tanks
- Piping systems (including piping components such as valves)
- Relief and vent systems and devices
- Emergency shutdown systems
- Controls including monitoring devices and sensors, alarms, and interlocks
- Pumps

Inspections and tests that follow recognized and generally accepted good engineering practices must be performed on process equipments. Frequency of inspections and tests of process equipments must be consistent with the manufacturers' recommendations and good engineering practices. Feedback from the prior operating experiences should become an integral part of the inspection methods. The employer must document each inspection and test that has been performed on process equipments, which contain the date of the inspection or test, name of the person who performed the inspection or test, serial number or other identification of the equipments that is inspected or tested, a description of the inspection or test performed and the results of the inspection or test conducted. Additionally, the employer must correct deficiencies in the equipments that do not meet the acceptable limits before they are put to use in the process line.

4.23 Management of Change

The employer must establish and implement written procedures to manage changes to process chemicals, technology, equipment, and procedures. Changes to the facilities that affect a covered process should also be indicated. The procedures must assure that the following considerations are addressed prior to any change:

- The technical basis for the proposed change;
- Impact of change on safety and health;

- Modifications to operating procedures;
- Necessary time period for the change;
- Authorization requirements for the proposed change.

Employees involved in the process operation including the contactor employees whose job tasks will be affected by the proposed change in the process must be informed well in advance. Employers should impart necessary training to all the employees involved in the specific process prior to the startup of the process or affected part of the process. If a change covered by the standard results in a change in the PSI required, such information must also be updated accordingly.

4.24 Incident Investigation

The employer must investigate as soon as possible, but not later than 48 hours following each incident that resulted in or could reasonably have resulted in a catastrophic release of a highly hazardous chemical in the workplace. An incident investigation team consisting of at least one person knowledgeable in the process involved, including a contract employee and other persons with appropriate knowledge and experience must be established to thoroughly investigate and analyze the incident. A report must be prepared at the conclusion of the investigation, which should include the date of the incident, date of investigation, description of the incident, factors that contributed to the incident, and recommendations resulting from the investigation. The employer must establish a system to promptly address and resolve the incident report findings and recommendations. Resolutions and corrective actions should be documented; such reports are retained for 5 years for record.

4.25 Compliance Audits

Employers must certify that they have evaluated compliance with the provisions of the Oil Industry Safety Directorate (OSID) standards at least every 3 years to verify that the procedures and practices developed under it are adequate and are being followed. The following must be adhered to in the compliance audit:

- It is conducted by at least one person knowledgeable in the process.
- A report of the findings of the audit is developed.

- The employer promptly determines and documents an appropriate response to each of the findings of the compliance audit, and documents that deficiencies have been corrected.
- Employers retain the two most recent compliance audit reports.

Unoccupied remote facility are those that are operated, maintained, or serviced by employees who visit the facility only periodically to check its operation and maintenance tasks; no employees are permanently stationed at such facilities. Facilities meeting this definition are not contiguous with and must be geographically remote from all other buildings, processes, or persons.

4.26 Software Used in HSE Management

Using integrated data capturing, advanced reporting, and trending on a software platform, risks and costs can be massively reduced while performance of the offshore assets can be improved. A few of the commercially available software, which includes most of the features in Health, Safety, and Environment (HSE) management are discussed in the following sections.

4.26.1 CMO Compliance

CMO Compliance is one of the commonly used software in HSE management. Leading oil industries around the world use this software to manage HSE processes and drive continual operational performance while reducing risks and costs. Faced with an increasingly regulated business environment with stricter laws and regulations, it is important that the process be supported by a good and updated HSE management system. This is helpful to ensure that the processes and practices followed are safe and supported by international safety regulations. A key step to achieve the workforce participation is a simple yet powerful health and safety management software tool for logging and managing incidents, scheduling activities, and managing actions. CMO Compliance is easy to access, use, and provide valuable real-time reports to team members that implement and manage the HSE processes. This software is designed to meet the requirements of HSE management team while driving the workforce to adopt the culture of participation in HSE for achieving the corporate goals.

4.26.2 Spiramid's HSE Software

Spiramid provides an integrated HSE software solution that addresses the trends in hazard management and risk analysis by utilizing HSE software reporting tools. Any implementation of changes compiled by the software will be able to play a large role in compliance, minimization of risk, as well as savings for business.

4.26.3 Integrum

Integrum, the global leading software for HSE quality and risk management is ideally suitable for the offshore oil and gas industries. Integrum is a web-based software that contains features such as observation cards, incident management and corrective actions, centralized controlled document management system, risk assessments and controls, audit, training management, competency assessments, contractor and supplier management, and consolidated management reporting. All these functions, combined in one application, allow Integrum's offshore oil and gas clients to quickly build, deploy, and manage HSE in a more comprehensive manner. The robust security and strong replication function of Integrum ensures that the controlled integrated management system can be easily replicated to remote locations, such as offshore vessels, where Internet connections are unreliable. The functionality of Integrum ensures that only changes are replicated, providing a fast reliable service for remote locations.

4.26.4 Rivo HSE Management Software

Rivo provides a total safety management software platform that is used by many leading oil and gas industries in the recent past. Rivo includes features such as audit management, corrective/preventive actions, environmental management, environmental risk assessment, forms management, incident management, industrial safety management, injury reporting, inspection management, Material Safety Data Sheet (MSDS), occupational health management, OSHA recordkeeping, safety risk assessment, training management, and waste management. A wide coverage of modules is very useful in managing HSE issues in a comprehensive manner.

Exercises 4

1. Substances causing .... properties result in oxygen deficiency.

   Saprogenic

2. Most dangerous type of plume is..............

   Fumigation

3. When inversion exists both below and above slack height .... results.

Trapping

4. When the earth's surface cools rapidly, such as between late night and early morning under clear skies, a .... inversion is likely to occur.

Radiation

5. .... refers to most unstable condition.

Stability class A

6. What are the different ways of water pollution that arise from offshore industry?

Drilling waste, oil spills, oil sludge, drilling solid waste, production waste.

7. What is the primary environmental issue that arises from oil and gas production?

Impact on the shelf ecosystems and marine biological resources contribute to the life hierarchy at different levels. They also significantly influence fishing. Biological consequences of accidental oil spills into the marine environment are irreversible.

8. What are the impacts on environment due to oil and gas production?

Geological and geographical survey: Interference with fisheries, impact on water organisms
Exploration: Discharge of pollution, interference with fisheries
Development and production: Operational discharges, accident spillage, physical disturbances
Decommissioning: Operational discharges, residual remains of the platform, impact on organisms when explosives are used.

9. Discuss the anthropogenic impact on hydrosphere with respect to environmental management in oil and gas sector.

Obvious or hidden disturbances of natural structure and function of water communities, changes in composition and characteristics of biotopes, alterations in hydrological regime and geomorphology of water bodies, diminishing fisheries, diminishing recreational values. Results in other ecological, economical, and socioeconomical consequences.

10. What are the consequences of marine pollutants?

    Oil slicks, tar balls, suspended solids, oil hydrocarbons: crude oil and oil products, hydrocarbons of methane series.

11. How produced waters during drilling cause marine pollution?

    Produced waters contain dissolved salts and organic compounds, oil hydrocarbons, trace metals, and suspensions are also present. Hence composition of produced water is very complex, generally it contains benzene, toluene, and xylenes (10–30 mg/kg in total), biocides (few mg/kg), organic molecules, and heavy metals. Chromatographic analysis of discharged water in Gulf of Mexico (GoM) showed very high and relatively stable levels of phenol and its alkylated homologues in the discharges. Even radio-active elements like radium-226 and radium-228 are seen in produced waters (GESAMP, 1991).

12. List the various factors that contribute to the estimate of consequence of marine pollutants?

    Hazardous properties of the pollutants, volume of their input into ocean, scale of distribution, pattern of their behavior in ecosystems, stability of their composition.

13. What are the different stages in oil and gas development?

    Geological and geographical survey, exploration, development and production, decommissioning.

14. What are the main constituents of oil-based drilling fluid?

    Barite: 409 tons (61%), base oil: 210 tons (31%), calcium chloride: 22 tons (3.35), emulsifier: 15 tons (2.2%), filtrate agent: 12 tons (1.8%), lime: 2 tons (0.25%), viscosifier: 2 tons (0.4%).

15. Write a brief note on drilling operations and their consequences.

    Drilling mud discharge is about 15–30 tons from a single well periodically, cuttings containing dry mass is about 200–1000 tons from a single well. In case of multiple wells, drilling mud is 45000 tons for about 50 wells, cutting is about 50000 tons for about 50 wells, waste discharge is about 1500 tons/day from a single production platform, volume of discharge in ocean in different parts of the world are very significant.

16. What do you understand by marine pollution? Why is it significant?

Large and multiscale activity of offshore oil and gas industry impose a complex impact on the marine environment. The impacts of marine pollution are chemical, physical, and biological in nature. Seismic signals generated during marine surveys are hazardous for marine fauna. Explosive activities of abandoned platforms result in mass migration of commercial fish. Chemical pollution is one of the most important impacts, large offshore accidents cause oil spills leading to serious ecological consequences. Fate of unused oil platforms and underwater pipelines cause serious threat to marine ecology.

17. Explain the fate and behavior of an oil spill.

Oil spill undergoes various stages, each stage pollutes marine environment significantly, such as physical transport, dissolution, emulsification, oxidation, sedimentation, microbial degradation, aggregation, and self-purification. Within 10 minutes of the spill of 1 ton of oil, oil can disperse over a radius of 50 m, it forms a thin slick of 10 mm thick. The slick gets thinner as oil continues to spread further. The area of spread of 1 ton oil spill can even extend as high as 12 km^2. During the first several days after the spill, a considerable part transforms into gaseous phase, slick gradually loses its water-soluble hydrocarbons, the remaining fraction, being viscous, reduces slick spreading.

18. What are the most common types of drilling discharges that take place during drilling?

Drilling muds are hazardous due to their persistence in marine environment, after 6 months of discharge of oil-based drilling waste, they biodegrade by only 5% (Ostgaard and Jensen, 1983). Drilling waste based on fatty acids lose their organic fraction due to microbial and physiochemical decomposition. Water-based drilling muds show higher dilution capacity in marine environment. Large volumes of water-based muds are disposed overboard which adds to marine pollution. Drill cuttings, which are pieces of rock crushed by drill bit and brought to surface do not pose any special threat.

19. What are the factors considered in the environment management policy making?

Balance of current and future interest, possibilities of alternative sources of energy, natural conditions, ecological, technical, and economical factors.

20. What are the important regulatory measures that have to be considered for discharging of drilling waste into sea?

Discharges into sea require authorization and must comply with regulations. Concentration of oil and oil products, determined using standard tests, should exceed established standards, LC_{50} values for discharge samples during 96-hour Mysid toxicity testing should not exceed 30 g/kg.

21. What are the different stages of ecological monitoring?

First, possible potential hazards from impact sources are identified, then regular observations of marine biota are conducted to qualitatively assess biological responses in organisms. In the next stage, cause–effect relationship between documented biological effects and impact factors are studied. The next stage is to assess the total impact on the marine environment and biota including the impact on commercial species and biological resources in general. Finally, corrective measures are incorporated for checking the marine pollution and preventive measures, if any.

22. How oil spills cause damage to the environment?

The effects of an oil spill depend on a variety of factors including the quantity and type of oil spilled, and how it interacts with the marine environment. Prevailing weather conditions also influence the oil's physical characteristics and its behavior. Other key factors include the biological and ecological attributes of the area, the ecological significance of key species and their sensitivity to oil pollution as well as the time of year. It is important to remember that the clean-up techniques selected will also have a bearing on the environmental effects of a spill.

23. How do oil spills occur?

When oil tankers have equipment faults, from nature and human activities on land, water Sports, drilling works carried out in sea.

24. What is meant by dispersion modeling?

It is an attempt to describe the relationship between emission, occurring concentration, and deposition. It gives a complete analysis of what emission sources have lead to concentration depositions, mathematical models use analytical and numerical formulations, usually implemented on computers.

25. What is the importance of dispersion modeling?

To predict ambient air concentration which results from an emission source, to plan and execute air pollution control program considering cost effectiveness. For environmental impact assessment, quantify the impact of process improvements, evaluate the performance of emission control techniques, optimize stack height, diameter, and plan the control of air pollution episodes.

26. What is meant by maximum mixing depth?

The depth of the convective mixing layer in which vertical movement of pollutants is possible is called the Maximum Mixing Depth (MMD).

27. What are the different types of inversions?

Subsidence inversion, radiation inversion, combination of subsidence and radiation.

28. What are the different ways of evaluating the toxicity of chemical releases?

PEL or TLV-TWAs, based on emergency response planning, guidelines as recommended by National Institute for Occupational Safety and Health (NIOSH), Guidelines as recommended by National Research Council, Canada (NRC).

Application Problem: Quantified Risk Assessment of LPG Filling Station

Introduction

Increase in the use of petrochemicals has led to more accidents with huge losses (Che Hassan et al., 2009, 2010). To minimize the consequences of such accidents within acceptable risk levels, strict codes of conduct and preventive policies are enforced by various administrative authorities (see, e.g., Oil Industry Safety Directorate [OISD]; Health and Safety Executive [HSE]; Occupational Safety and Health Administration [OSHA] and so on). Given the fact that accidents (or near-misses) are inevitable in oil and gas industries, it is a common practice to carry out the risk assessment for such scenarios to ensure safe working practices inside the plant. Common hazards posed by the LPG filling station are dispersion, jet fire, fireball, and BLEVE (Vanem et al., 2008). Risks associated with such hazards are also classified, but no quantitative studies are reported (Pontiggia et al., 2011). The current study facilitates a better idea of hazards, risks, and consequences involved with the LPG filling station (Chandrasekaran and

Health, Safety, and Environmental Management in Offshore and Petroleum Engineering, First Edition.
Srinivasan Chandrasekaran.
© 2016 John Wiley & Sons, Ltd. Published 2016 by John Wiley & Sons, Ltd.
Companion website: www.wiley.com/go/chandrasekaran/hse

Kiran, 2015). Detailed risk analyses are conducted using Det Norske Veritas (DNV) Phast Risk software.

Some of the consequences of the release of LPG are dispersion, jet fire, fireball, and boiling liquid expanding vapor explosion (Gomez-Mares et al., 2008; Zhang and Liang, 2013).

Dispersion

Dispersion is the accidental discharge of flammable or toxic materials as pressurized liquid, gas, or vapor. Greater hazard will be generally due to the release of pressurized liquid discharge. In the present study, dispersion effect is important in calculating the lower flammability region of LPG release. Lower flammability region is the region in which the fuel will not get ignited below this concentration. The Lower Flammability Limit (LFL) region for LPG with 60% butane and 40% propane is computed as 16 999 ppm.

Jet Fire

Jet fire is an intense, highly directional fire resulting from the ignition of a vapor or two-phase release with significant momentum. A jet fire is a result of combustion and ignition of a flammable fluid releasing from a pipe or an orifice. Jet fires cause thermal radiation, which will transmit heat energy causing damage to nearby properties and fatalities to the workers in the plant.

Fireball

Fireballs are due to ignition of turbulent vapor or two-phase fuel in air with a short duration. Fireballs are instantaneous in nature and are generally due to catastrophic failure of pressurized vessels. Fireballs produce large amount of thermal radiation, which will transmit heat energy to the surroundings.

Boiling Liquid Expanding Vapor Explosion (BLEVE)

Boiling liquid expanding vapor explosion is due to the sudden loss of containments above its normal boiling point at the time of vessel failure. This results in the development of cracks, which may be due to fire engulfment of a vessel containing liquid under pressure. Due to fire outside the vessel, the liquid inside gets vaporized. This subsequently activates the safety valve, which increases the vapor content inside the pressure vessel. Nonuniform expansion of the wall of the vessel takes place as the heat capacity of the vapor is lesser than that of the liquid. This causes loss of strength and release of containments suddenly. The sudden release produces shock waves, which damages the plant and also causes fatalities.

Risk is defined as the probability of occurrence of events and its consequences. It can be expressed in terms of individual risk and societal risk. Individual risk is the frequency at which the individual may be expected to sustain a given level of harm from the realization of hazard. It is the ratio of number of fatalities and number of people at risk. Societal risk can be defined as the relation between frequency and number of people suffering from the realization of hazard. Societal risks are generally expressed as frequency-number of fatality curve (F-N curve).

Methodology

For the case study two LPG filling stations at different locations are considered. It is interesting to note that no real accident or near-miss took place in these plants; accident scenarios are pseudo- created for the study. Layouts of both the plants are given in Figures 1 and 2, respectively. Relevant failure cases and their respective consequences are given in Table 1. Input data required for the analysis are chemical properties of LPG, different release scenarios, in-situ storage conditions, and weather data. For the analysis, an average value of the weather conditions is taken for the year 2013 (Indian Meteorological Department). During the site inspection of the plants, it is noted that LPG is stored in the pressurized vessel at 5–7 kg/cm². Based on the listed input conditions, hazard distances for these failure scenarios are determined from the numerical analysis using the software. For the present study, the consequence assessment is carried out in terms of quantified hazard distances. Risk is calculated for these consequences in terms of individual and societal risk. The probability of failure of different failure cases is taken from the handbook. The calculated risk is then compared with that of the safety standards to check whether these are within the permissible limits. Safety measures are recommended for the cases that are not within the permissible limits and then the risk is recalculated with these updated safety measures.

Results and Discussions

Dispersion

Hazard distances for different failure cases are give in Table 2. It can be seen that the maximum LFL hazard distance is 157 m for Plant A and 101 m for Plant B is due to the catastrophic failure of the storage bullet. It is also seen that there is a significant increase in the hazard distance with the increase in the mass.

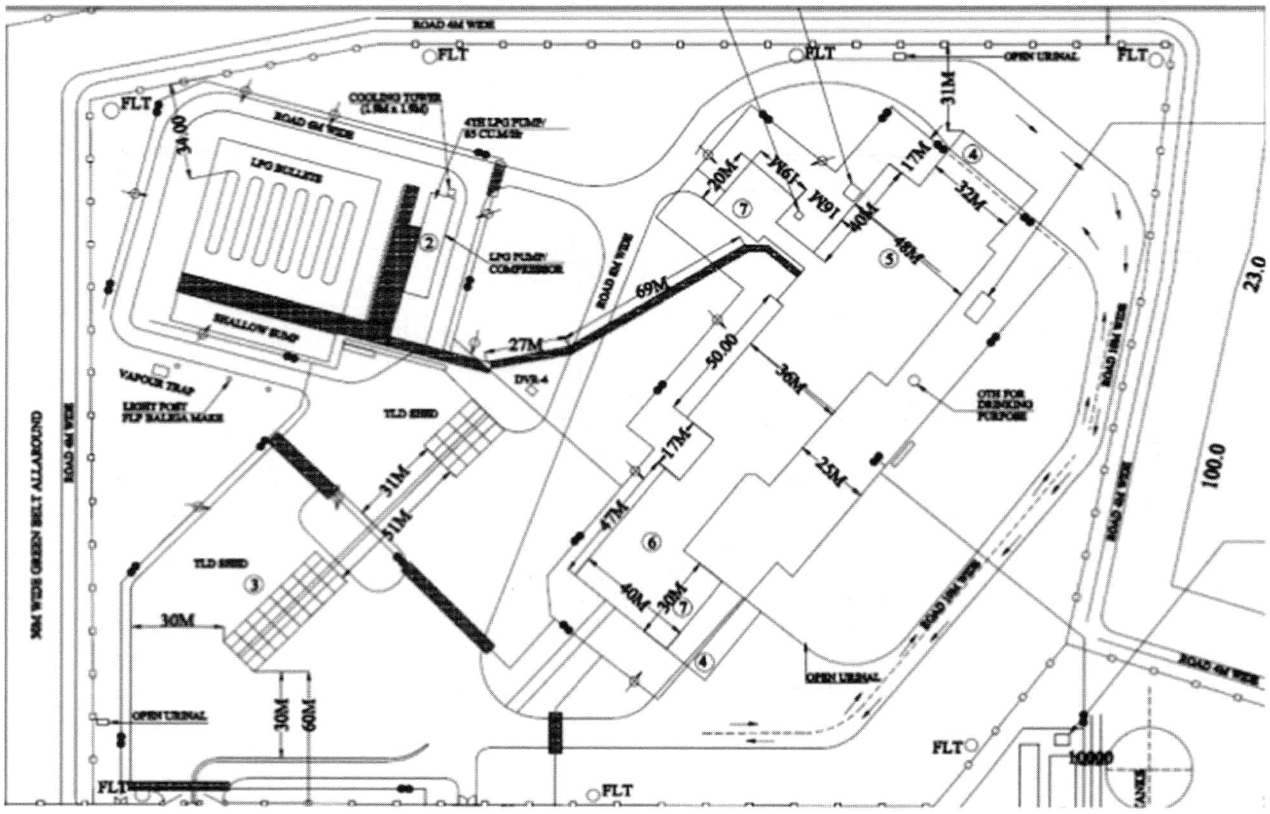

Figure 1 Layout of Plant A

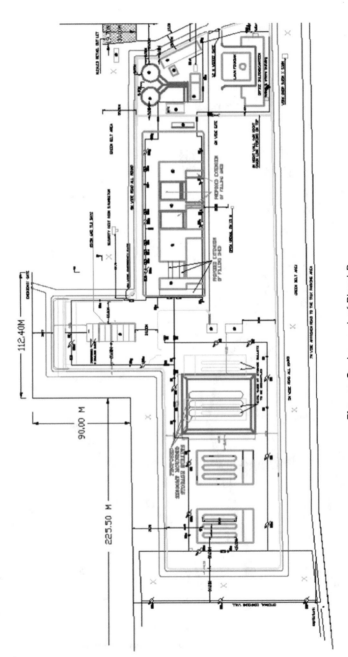

Figure 2 Layout of Plant B

Table 1 Failure cases and their consequences

Sl. no.	Failure case	Consequences
1.	Full bore failure of LPG outlet line of bullets	Dispersion, jet fire
2.	20% cross-sectional area (CSA) failure of LPG outlet line of bullets	Dispersion, jet fire
3.	LPG pump discharge line full bore failure	Dispersion, jet fire
4.	Road tanker failure	Dispersion, fireball, BLEVE
5.	LPG pump mechanical seal failure	Dispersion, jet fire
6.	LPG pump outlet line gasket failure	Dispersion, jet fire
7.	Road tanker unloading arm failure	Dispersion, jet fire
8.	Catastrophic failure of a single bullet	Dispersion, fireball, BLEVE
9.	LPG unloading vapor compressor outlet line full bore failure	Dispersion, jet fire

Table 2 Dispersion distances

Sl. no.	Failure case	LFL hazard distance for Plant A (m)	LFL hazard distance for Plant B (m)
1.	Full bore failure of LPG outlet line of bullets	67	67
2.	20% CSA failure of LPG outlet line of bullets	24	24
3.	LPG pump discharge line full bore failure	40	40
4.	Road tanker failure	139	71
5.	LPG pump mechanical seal failure	28	29
6.	LPG pump outlet line gasket failure	33	33
7.	Road tanker unloading arm failure	21	23
8.	Catastrophic failure of a single bullet	157	101

Thermal Radiation Due To Jet Fire

Jet fire also resulted in an increase in the thermal radiation. For higher intensity of thermal loads, it is seen that there is an increase in the damage to the plant and equipments. It is therefore important to know these hazard distances to

Table 3　Hazard distance due to jet fire

Sl. no.	Failure case	Hazard distance for intensity load 37.5 kW/m² (Plant A)	Hazard distance for intensity load 37.5 kW/m² (Plant B)
1.	Full bore failure of LPG outlet line of bullets	54	50
2.	20% CSA failure of LPG outlet line of bullets	28	25
3.	LPG pump discharge line full bore failure	36	39
4.	LPG pump mechanical seal failure	29	30
5.	LPG pump outlet line gasket failure	32	35
6.	Road tanker unloading arm failure	24	26

Table 4　Hazard distances due to fireball

Sl. no.	Failure scenario	Hazard distance for intensity load 12.5 kW/m² (Plant A) (m)	Hazard distance for intensity load 12.5 kW/m² (Plant B) (m)
1.	Road tanker failure	187	189
2.	Catastrophic failure of a single bullet (capacity: 150 MT)	371	375

keep the critical equipments away from the said source. The hazard distances due to the thermal radiation from jet fire intensity of 37.5 kW/m² thermal load is calculated for various scenarios and given in Table 3.

Thermal Radiation Due To Fireball

Fireballs are generally short-lived flames, as discussed earlier, and hence do not cause a thermal load as high as 37.5 kW/m². Due to the catastrophic failure of the storage bullet, it is found that an intensity of 12.5 kW/m² thermal load produced a hazard distance of 371 m for Plant A and 375 m for Plant B. The hazard distance from fireball due to different failure cases is given in Table 4.

Table 5 Hazard distances due to overpressure from BLEVE

Sl. no.	Failure scenario	Hazard distance for intensity load of 0.3 bar (Plant A) (m)	Hazard distance for intensity load of 0.3 bar (Plant B) (m)
1.	Road tanker failure	58	58
2.	Catastrophic failure of a single bullet (capacity: 150 MT)	129	129

Overpressure Effects Due To BLEVE

Due to BLEVE shock waves are generated. Generally when an explosion occurs, major damage is caused due to shock waves rather than due to thermal radiation. It is important to note that an intensity of 0.3 bar shock waves is sufficient enough to damage the plant (OISD). Due to the catastrophic failure of storage bullet, as envisaged in the study, hazard distance is computed as 129 m for both the plants. The hazard distances due to overpressure from BLEVE is given in Table 5.

Risk Estimates

For different consequences, individual and societal risks are calculated for various failure scenarios and these are given in Table 6. It can be seen that the risk is higher for catastrophic failure of storage pressurized bullets. For Plant A, the risk is found to be 1.1E−4 and 3.3E−5 per average year for catastrophic failure of storage bullets in Plants A and B, respectively. Similarly, for the road tanker failure, risk is found to be 1.2E−5 and 9.1E−6 per average year for Plants A and B, respectively. For road tanker unloading arm failure, individual risk is found to be 3.6E−5 and 2.7E−5 per average year for Plants A and B, respectively. As discussed earlier, failure scenarios are seen as higher risk events in comparison to the other failure scenarios for both the plants.

The risk is then compared with that of the permissible standards as shown in Figure 3, which is in the form of an ALARP (as low as reasonably practical) triangle. It is seen from the figure that the acceptable risk for existing hazardous industries is 1E−6 per average year and the intolerable risk is 1E−4 per average year. The calculated risk is compared with that of the acceptable ones. It is seen that for some of the failure cases like catastrophic failures of storage bullets and road tankers, it is not within the acceptable limits. For Plant A, due

Table 6 Individual and societal risks for different failure cases

Sl. no.	Failure scenario	Plant A		Plant B	
		Individual risk (per average year)	Societal risk (per average year)	Individual risk (per average year)	Societal risk (per average year)
1.	Full bore failure of LPG outlet line of bullets	2.5E–008	1.7E–008	2.4E–008	2.2E–008
2.	20% CSA failure of LPG outlet line of bullets	8.5E–009	5.6E–009	8.2E–009	5.8E–009
3.	Catastrophic failure of storage bullets	1.1E–004	7.4E–005	4.4E–005	3.3E–005
4.	Road tanker failure	1.2E–005	8.7E–006	9.1E–006	8.5E–006
5.	LPG pump discharge line full bore failure	2.4E–008	1.8E–008	5.4E–007	4.9E–007
6.	LPG pump outlet line gasket failure	2.5E–008	1.9E–008	4.1E–007	3.7E–007
7.	Road tanker unloading arm failure	3.6E–005	2.2E–005	2.7E–005	2.2E–005
8.	Vapor compressor line failure	9.1E–008	5.5E–008	9.5E–008	7E–008

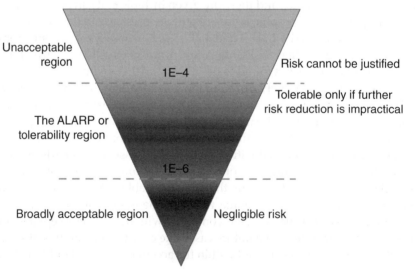

Figure 3 ALARP triangle

Table 7 Risks after recalculations

Failure scenario	Plant A		Plant B	
	Individual risk (per average year)	Societal risk (per average year)	Individual risk (per average year)	Societal risk (per average year)
Catastrophic failure of storage bullets	6.1E–8	6.3E–8	6.1E–8	6.3E–8
Road tanker failure	5.2E–7	3E–7	3.7E–7	3E–7
Road tanker unloading arm failure	9E–7	5.4E–7	5.3E–7	4.1E–7

to catastrophic failure of storage bullets, the risk is unacceptable and for the Plant B it is in ALARP region. Therefore, to reduce the risk and consequences, mounded storage bullets are recommended instead of unmounded storage bullets. In the road tanker bays, personnel concentration should be avoided. Further, battery of the tanker should be disconnected and proper earth should be provided while loading and unloading. Assuming strict implementation of the recommended safety measures, risk is recalculated and given in Table 7.

Conclusions

The consequence analysis and risk assessment of LPG filling stations located at two different places are discussed. This work is a preliminary study in risk assessment of LPG installations, which helps in expansion or installation of new equipments in the existing plant or part thereof. Based on the pseudo accident scenarios assumed for the both the plants, it was found that the risk involved for the catastrophic failure for Plants A and B are not within the acceptable risk of 1E–6 per average year. But, after the recommendation of converting into mounded bullets, the risk is reduced to 6.1E–8 per average year for both the plants. For road tankers' failure, risk was found to be in the ALARP region. But this is reduced by implementing proper safety measures such as removing the connection from the battery, providing earth etc. After such measures, the risk is found to be reduced to 5.2E–7 per average year for Plant A and 3E–7 per average year for Plant B, which is within the permissible limits. The current study gives a better idea of risk involved for different failure scenarios associated with LPG filling stations. The study can be very helpful in deciding expansion of the plant or installation of new plant at a nearby locality.

References

Abbasi T. and S. A. Abbasi (2007) The boiling liquid expanding vapor explosion (BLEVE): mechanism, consequence assessment and management, Journal of Hazardous Materials, 141, 489–519.

Ale B. J. M. (2002) Risk assessment practices in the Netherlands, Safety Science, 40, 105–126.

Amendola A., S. Continin, I. Ziomas (1992) Uncertainties in chemical risk assessments: results of a European benchmark exercise, Journal of Hazardous Materials, 29, 347–363.

Arshad Ayub (2011) *Hazard identification and management in oil and gas industries,* M.Tech (Petroleum Engg) dissertation submitted to IIT Madras.

Aven T. and J. E. Vinnem (2007) *Risk Management with Applications from Offshore Petroleum Industry,* Springer, London, 200pp.

Bhattacharyya S. K., S. Chandrasekaran, R. Prasad (2010a) Event analysis for offshore riser failure. Proceedings of the First International Conference on Drilling Technology (ICDT 2010), IIT Madras, India, November 18–20, 2010.

Bhattacharyya, S. K., S. Chandrasekaran, R. Prasad (2010b) Risk assessment for off-shore pipelines. Proceedings of the First International Conference on Drilling Technology (ICDT 2010), IIT Madras, India, November 18–20, 2010.

Bonvicini S., P. Leonelli, G. Spadoni (1998) Risk analysis of hazardous materials transportation: evaluating uncertainty by means of fuzzy logic, Journal of Hazardous Materials, 62, 59–74.

Bottelberghs P. H. (2000) Risk analysis and safety policy developments in the Netherlands, Journal of Hazardous Materials, 71, 59–84.

Brazier A. M. and R. L. Greenwood (1998) Geographic information systems: a consistent approach to land use planning decisions around hazardous installations, Journal of Hazardous Materials, 61, 355–361.

Health, Safety, and Environmental Management in Offshore and Petroleum Engineering, First Edition.
Srinivasan Chandrasekaran.
© 2016 John Wiley & Sons, Ltd. Published 2016 by John Wiley & Sons, Ltd.
Companion website: www.wiley.com/go/chandrasekaran/hse

Brode H. L. (1959) Blast wave from a spherical charge, Physics of Fluids, 2(2), 217–229.

Bubbico R. and M. Marchini (2008) Assessment of an explosive LPG release accident: a case study, Journal of Hazardous Materials, 155(3), 558–565.

Cairns W. J. (Ed) (1992) *North Sea Oil and the Environment: Development Oil and Gas Resources, Environmental Impacts and Responses*, International Council of Oil and the Environment, Elsevier, London.

Cano M. L. and P. B. Dorn (1996) Sorption of two model alcohol ethoxylate surfactants to sediments, Chemosphere, 33, 981–994.

Chamberlain G. A. (1987) Development in design methods for predicting thermal radiation from flares, Chemical Engineering Research and Design, 65, 299–309.

Chandrasekaran S. (2010a) Chemical risks—an overview, Key note address at HSE in Oil and gas-exploration and production, International HSE Meet, IBC-Asia, Kuala Lumpur, Malaysia, December 6–8, 2010.

Chandrasekaran S. (2010b) Risk assessment of offshore pipelines, Key note address at HSE in Oil and gas-exploration and production, International HSE Meet, IBC-Asia, Kuala Lumpur, Malaysia, December 6–8, 2010.

Chandrasekaran S. (2011a) Hazard identification and management in oil and gas industry using Hazop. Proceedings of seminar on Human Resource Development for Offshore and Plant Engineering (HOPE), Changwon University, South Korea, April 2011, pp. 1–10.

Chandrasekaran S. (2011b) Health, Safety and Environmental Management in petroleum and offshore engineering. Proceedings of seminar on Human Resource Development for Offshore and Plant Engineering (HOPE), Changwon University, South Korea, April 2011, pp. 23–28.

Chandrasekaran S. (2011c) Quantitative risk assessment of Group Gathering Station (GGS) of oil exploration and production. Proceedings of seminar on Human Resource Development for Offshore and Plant Engineering (HOPE), Changwon University, South Korea, April 2011, pp. 11–22.

Chandrasekaran S. (2011d) Strategic Rig Project Commissioning and Risk Management, Keynote address at Post-conference Workshop on International Conference on Offshore Drilling Rigs, IBC Asia, Singapore, July 24–29, 2011.

Chandrasekaran S. (2011e) Risk Assessment and Management in Offshore and Petroleum Industries, Key note address at Pre-conference workshop on International Conference on Asia Pacific HSE Forum on Oil, Gas and Petrochemicals, Fleming Gulf Conferences, September 22–24, 2011, Kuala Lumpur, Malaysia, pp. 81.

Chandrasekaran S. (2014a) Heath, Safety and Environmental Management (HSE), Key note address at National Research Foundation of Korea, HRD Team for Offshore Plant FEED Engineering, Changwon National University, South Korea, February 25, 2014.

Chandrasekaran S. (2014b) Technological Advancements in Process Safety Management, Key note address in the 4th Annual HSE Excellence Forum in oil, Gas and Petrochemicals, August 19–21, 2014, Kuala Lumpur, Malaysia.

Chandrasekaran S. (2015) HSE in offshore and petroleum engineering, Lecture notes of online web course, Mass Open-source Online Courses (MOOC), National Program on Technology Enhancement and Learning (NPTEL), Govt. of India.

Chandrasekaran S. and N. A. Harinder (2011) Design and efficiency analysis of mechanical wave energy converter. Proceedings of 30th International Conference on Ocean, Offshore and Arctic Engineering, OMAE 2011, Rotterdam, the Netherlands, June 19–24, 2011, OMAE 2011-49830.

Chandrasekaran S. and N. A. Harinder (2014) Failure mode and effects analysis of mechanical wave energy converters, International Journal of Intelligent Engineering Informatics, 3(1), 57.

Chandrasekaran S. and A. Kiran (2014a) Accident Modeling & Risk Assessment of Oil & Gas Industries. Proceedings of 9th Structural Engineering Convention (SEC 2014), IIT Delhi, India, December 22–24, 2014.

Chandrasekaran S. and A. Kiran (2014b) Consequence analysis and risk assessment of oil and gas industries. Proceedings of International Conference on Safety & Reliability of Ship, Offshore and Subsea Structures, Glasgow, UK, August 18–20, 2014.

Chandrasekaran S. and A. Kiran (2015) Quantified risk assessment of LPG filling station, *Professional Safety*, Journal of American Society of Safety Engineers (ASSE), September 2015, pp. 44–51.

Chandrasekaran S., Ramesh Babu, Arshad Ayub (2010) Hazop study for crude oil pipe line. Proceedings of 1st International Conference on Drilling Technology (ICDT 2010), IIT Madras, India, November 18–20, 2010.

Che Hassan C. R., B. Puvaneswaran, A. R. Aziz, M. Noor Zalina, F. C. Hung, N. M. Sulaiman (2009) A case study of consequences analysis of ammonia transportation by rail from Gurun to Port Klang in Malaysia using safety computer model, Journal of Safety Health and Environment, 6(1), Spring, 1–19.

Che Hassan C. R., B. Puvaneswaran, A. R. Aziz, M. Noor Zalina, F. C. Hung, N. M. Sulaiman (2010) Quantitative risk assessment for the transport of ammonia by rail, American Institute of Chemical Engineers Process Safety Progress, 29, 60–63.

Chuan-Jie Zhu, Bai-quan Lin, Bing-you Jiang, Qian Liu, Yi-du Hong (2013) Numerical simulation of blast wave oscillation effects on a premixed methane/air explosions in closed end ducts, Journal of Loss Prevention in the Process Industries, 26, 851–861.

Crawley F., M. Preston, B. Tyler (2000) *HAZOP: Guide to Best Practice. Guidelines to Best Practice for the Process and Chemical Industries.* European Process Safety Centre and Institution of Chemical Engineers, Rugby, Warwickshire, UK.

David Brown F. and E. William Dunn (2007) Application of a quantitative risk assessment method to emergency response planning, Computers & Operations Research, 34, 1243–1265.

DNV Phast Risk, Det Norske Veritas, 2005, User manual-Version 6.7.

Dziubinski M., M. Fratczak, A. S. Markowski (2006) Aspects of risk analysis associated with major failures of fuel pipelines, Journal of Loss Prevention in the Process Industries, 19, 399–400.

Engelhard W. F. J. M., F. H. de Klepper, D. W. Hartmann (1994) Hazard analysis for the Amoco Netherlands PI1S-PI1S production facilities in the North Sea. Proceedings of SPE International Conference on Health, Safety and Environment, Jakarta, January 25–27, 1994.

Frank K. H. and H. W. Morgan (1979) A logical risk process of risk analysis, *Professional Safety*, June, 23–30.

GESAMP (1991) *Global Strategies for Global Environmental Protection with Addendum*, International Maritime Organization, London, UK.

Gomez-Mares M., L. Zarate, J. Casal (2008) Jet fires and the domino effect, Fire Safety Journal, 43, 583–588.

Henselwood F. and G. Phillips (2006) A matrix based risk assessment approach for addressing linear hazards such as pipelines, Journal of Loss Prevention in the Process Industries, 19, 433–441.

IEC 61882, Hazard and Operability Studies (HAZOP Studies) – Application Guide, International Electro Technical Commission, Geneva.

IS1656:2006, Indian Standard Hazard Identification and Risk Analysis-Code of Practice, Bureau of Indian Standards, 2006.

Johnson D. W. and J. B. Cornwell (2007) Modeling the release, spreading and burning of LNG, LPG and gasoline on water, Journal of Hazardous Materials, 140(3), 535–540.

Khan F. I. and S. A. Abbasi (1999) Major accidents in process industries and analysis of causes and consequences, Journal of Loss Prevention in the Process Industries, 12(5), 361–378.

Kiran A. (2012) *Risk analyses of offshore drilling rigs*, M.Tech (Petroleum Engg) dissertation submitted to IIT Madras.

Kiran A. (2014) *Accident modeling and risk assessment of oil and gas industries*, M.S. (by research) thesis submitted to IIT Madras.

Kletz T. (2003) *Still Going Wrong: Case Histories and Plant Disasters*, Elsevier, Chennai, India, pp. 230.

Kyriakdis I. "HAZOP—Comprehensive Guide to HAZOP in CSIRO," CSIRO Minerals, National Safety Council of Australia, 2003.

Lees F. P. (1996) *Loss Prevention in Process Industries: Hazard identification, Assessment and Control*, Vol. 1–3, Butterwort-Heinemann, Oxford, 1245pp.

Leonelli P., S. Bonvicini, G. Spadoni (1999) New detailed numerical procedures for calculating risk measures in hazardous material transportation, Journal of Loss Prevention in the Process Industries, 12, 507–515.

Michailidou E. K., K. D. Antoniadis, M. J. Assael (2012) The 319 major industrial accidents since 1917, The International Review of Chemical Engineering, 4(6), 1755–2035.

Michailidou E. K., K. D. Antomiadis, M. J. Assael (2012) The 319 Major Industrial Accidents Since 1917, International Review of Chemical Engineering, 4(6), 1755–2035.

Nivolianitou Z., M. Konstandinidou, C. Michalis (2006) Statistical analysis of major accidents in petrochemical industry notified to the major accident reporting system (MARS), Journal of Hazardous Materials, 136(1), 1–7.

OGP Risk Assessment Data Directory: Report No.434-1, Process Release Frequencies, March 2010.

OISD - GDN - 169, OISD Guidelines on Small LPG Bottling Plants (Design and Fire Protection Facilities), Oil Industry Safety Directorate, Amended edition, 2011.

OISD Standard - 116, Fire Protection Facilities for Petroleum Refineries and Oil/Gas Processing Plants, Oil Industry Safety Directorate, Amended edition, 2002.

OISD Standard - 144, Liquefied Petroleum Gas (LPG) Installations, Oil Industry Safety Directorate, Second edition, 2005.

OISD Standard - 150, Design and Safety Requirements for Liquefied Petroleum Gas Mounded Storage Facility, Oil Industry Safety Directorate, 2013.

Ostgaard K. and A. Jensen (1983) Preparation of aqueous petroleum solution for toxicity testing, Environmental Science and Technology, 17, 548–553.

Papazoglou I. A., L. J. Bellamy, O. N. Aneziris, B. J. M. Ale, J. G. Post, J. I. H. Oh (2003) I-risk: development of an integrated technical and management risk methodology for chemical installations, Journal of Loss Prevention in the Process Industries, 16, 575–591.

Pasman H. J., S. Jung, K. Prem, W. J. Rogers, X. Yang (2009) Is risk analysis a useful tool for improving process safety, Journal of Loss Prevention in the Process Industries, 22, 769–777.

Patin S. (1999) *Environmental Impact of the Offshore Oil and Gas Industry*, Eco Monitor Publishing, East Northport, 425pp.

Planas-Cuchi E., J. M. Salla, J. Casal (2004) Calculating overpressure from BLEVE explosions, Journal of Loss Prevention in the Process Industries, 17, 431–436.

Pontiggia M., G. Landucci, V. Busini, M. Derudi, M. Alba, M. Scaioni, S. Bonvicini, V. Cozzani, R. Rota (2011) CFD model simulation of LPG dispersion in urban areas, Atmospheric Environment, 45(24), 3913–3923.

Prem K. P., D. Ng, M. S. Mannan (2010) Harnessing database resources for understanding profile of chemical process industry incidents, Journal of Loss Prevention in the Process Industries, 23(4), 549–560.

Qian-xi Zhang and Liang Dong (2013) Thermal radiation and impact assessment of the LNG BLEVE fireball, Procedia Engineering, 52, 602–606.

Ramamurthy K. (2011) *Explosions and Explosion Safety*, Tata McGraw Hill, New Delhi, India, pp. 288.

Rodante T. V. (2004) Analysis of an LPG explosion and fire, Process Safety Progress, 22, 174–181.

Skelton B. (1997) *Process Safety Analysis*, Gulf Publishing Company, Houston, 210pp.

Sutherland V. J. and C. L. Cooper (1991) *Stress and Accidents in Offshore, Oil and Gas Industries*, Gulf Publishing Co., Houston, pp. 227.

TNO (1999) Guidelines for quantitative risk analysis, The Director General of Labour, The Hague, the Netherlands.

Vanem E., P. Antao, I. Østivik, F. D. C. de Comas (2008) Analyzing the risk of LNG carrier operations, Reliability Engineering and System Safety, 93, 1328–1344.

Venkata Kiran G. (2011) *QRA in oil & gas industries using PHAST RISK*, M.Tech (Petroleum Engg) dissertation submitted to IIT Madras.

Vinnem J. E. (2007a) *Offshore Risk Assessment: Principles, Modeling and Applications of QRA Studies.* Springer, London, 577pp.

Vinnem J. E. (2007b) *Offshore Risk Assessment: Principles, Modeling and Applications of QRA* (2nd Ed.). Springer, New York.

Webber D. M., S. J. Jones, G. A. Tickle, T. Wren (1992) A Model of a Dispersing Gas Cloud, and the Computer Implementation. I: Near Instantaneous Release, II: Steady Continuous Releases. UKAEA Reports SRD/HSE R586 (for part I) and R. 587 (for part II).

Wiltox H. W. M. (2001) Unified Dispersion Model (UDM), Theory Manual, Det Norske Veritas, Houston, TX.

Index

Page numbers in *italics* refer to figures and/or tables. The index is organized in letter-by-letter order, i.e. spaces and hyphens are ignored in the alphabetical sequence.

Health, Safety, and Environmental Management in Offshore and Petroleum Engineering, First Edition.
Srinivasan Chandrasekaran.
© 2016 John Wiley & Sons, Ltd. Published 2016 by John Wiley & Sons, Ltd.
Companion website: www.wiley.com/go/chandrasekaran/hse